GOOD NEWS, PLANET EARTH!

GOOD NEWS, PLANET EARTH!

WHAT'S BEING DONE TO SAVE OUR WORLD, AND WHAT YOU CAN DO TOO!

SAM BENTLEY

Publisher Mike Sanders
Senior Editor Alexander Rigby
Art & Design Director William Thomas
Designer Lindsay Dobbs
Illustrator Cindy Kang
Photo Researcher Micah Schmidt
Proofreaders Mira Park, Monica Stone
Fact Checker Devon Fredericksen
Indexer Johnna VanHoose Dinse

First American Edition, 2023
Published in the United States by DK Publishing
DK, a Division of Penguin Random House LLC
6081 E. 82nd Street, Suite 400, Indianapolis, Indiana 46250

The authorized representative in the EEA is Dorling Kindersley
Verlag GmbH. Arnulfstr. 124, 80636 Munich, Germany

Library of Congress Catalog Number: 2022950523
ISBN 978-0-7440-8158-9

DK books are available at special discounts when purchased in bulk for sales promotions, premiums, fund-raising, or educational use.
For details, contact: SpecialSales@dk.com

Printed and bound in China

For the curious
www.dk.com

TO EVERYONE FIGHTING TO CREATE
A BETTER PLANET

PART 1: OCEANS 14

PART 2: LAND 36

PART 3: FOOD 66

PART 4: WILDLIFE 100

PART 5: CITY 126

PART 6: ENERGY 156

PART 7: TRANSPORTATION 186

INTRODUCTION

How did I find myself here, writing a book about climate action and all the good things that are happening on our planet? Let's rewind a little bit.

Although you wouldn't guess it from my accent, I was born in the great city of Birmingham in England. We Brummies—the name for people born in Birmingham—have a very distinct accent that is often mocked by others! And for most of my life, I didn't really care about the planet.

Well, it's not that I didn't care. I've always loved animals and stunning natural landscapes, but it wasn't something I really thought too deeply about. I loved watching football, creating art and music, and trying my hand at different entrepreneurial ventures.

During secondary school, I sold Yu-Gi-Oh! cards, an EP I created with four songs I wrote (mainly love songs because I'm a sensitive soul), and DVDs of movies I ripped from LimeWire—RIP to my parents' old computer. And in my last couple years at school, I started selling street art to my friends, which I was producing for my A-levels (college-entrance exams). This was really the start of my entrepreneurial journey as an adult.

I went to the University of Nottingham to study architecture, something I'd heard was a respectable career option, but not one I connected with very deeply. I mainly went because during the second year, the architecture students got to design and build a school in South Africa. Before coming to uni (college), I'd visited South Africa twice on volunteering trips, and I absolutely loved it and wanted to return.

While I was trying my best to succeed at architecture and learning how much alcohol my liver could withstand, I spent my free time growing the street art business I'd started at the end of secondary school. I created stencils of popular figures, spray painted them onto T-shirts and canvases, and sold them through a website my friend helped me create (who has now sadly passed away due to brain cancer). I didn't make much from it, but the satisfaction of creating something and seeing it out there in the world felt amazing.

My street art business went on to win the Creative Industries Award at the NACUE (National Association of College & University Entrepreneurs) Varsity Pitch competition, which is the UK's biggest business competition for students and graduates. I also ended up representing the University of Nottingham at the Santander X Entrepreneurship Awards. I didn't do so well with my architecture courses however, getting a 2.2.

It was at this point I was presented with a new opportunity.

The Varsity Pitch competition was held in London, and I had nowhere to stay while there. My housemate in Nottingham introduced me to his best friend, Liam, who lived there and said I could stay with him.

Liam was involved with a publisher who had gained some momentum. The publisher was called UNILAD, which catered to lads. (*Lads* are "young men," if you're not familiar with the British term.) While UNILAD wasn't producing content that resonated with me at the time, I saw massive potential to build a platform that could help create positive change, moving away from lad content and toward something more meaningful.

I shared my ideas, did a short trial with Liam to see if I'd be a valuable addition to the team, and became part of UNILAD. Oh, and I also dropped out of my fourth year of architecture school at this point—no surprises there.

Over the course of the next six years, with the help of an incredibly talented team, we grew a small Facebook page into one of the biggest social media companies in the world with over 60 million fans and over 1 billion video views every month.

I became obsessed with social media, consistently looking for new ways to pioneer how content is digested and shared. I wanted to keep UNILAD at the forefront of the social media industry.

As a reward for this obsession, Liam and I were awarded a place on the *Forbes* "30 Under 30" list in 2017, which was pretty awesome to say the least.

In between the viral videos of cats and dogs, we created content around meaningful topics, partnering with organizations like Campaign Against Living Miserably, World Animal Protection, and various homeless charities.

One of the content series I became most proud of during my time there was a series I created called *The Dark Truth Behind Tourism*, which exposed the horrendous ways animals are treated in the tourism industry, such as elephant riding and having photos taken with dolphins.

In some instances, shortly after posting an example of animal abuse that got massive exposure, we had local authorities in that region formally announce a plan to change how these activities operated to be more compassionate toward the animals involved.

This was when I truly realized the power social media can have to foster good.

I moved on from UNILAD in 2018 and, using my knowledge of social media, turned my focus to creating content around how we can create a more sustainable and equitable future. No more sourcing funny cat videos, unfortunately.

On this journey, I became aware of some of the huge issues our planet is facing. Documentaries like *Before the Flood* with Leonardo DiCaprio exposed me to the damage the beef industry is having on our planet and got me questioning how my own consumption affects the environment and what I could do to be part of the solution.

These topics seemed overwhelming, and many were hard to digest. But through countless hours of tailoring content for social media and making them as accessible and to the point as possible, I became an expert in translating hard-to-understand topics into short, digestible videos. With this experience, I knew I could become part of the solution.

After a few years of consulting for companies and navigating the pandemic, I decided to take the leap and launch my own social media platforms, using all the skills I'd used to build UNILAD and other platforms by applying them to my own channels.

Amongst all the doom and gloom news you see when you click on the TV or check social media, I wanted to bring a light to the good news happening on our planet and present new ways anyone can get involved in climate action.

SO WHY AM I WRITING THIS BOOK?

I am lucky enough to reach millions of people every month with my content online, but there are millions of people out there who aren't on social media who want to know about the good things happening and how they can help. That's where this book comes in.

One of the most frequently asked questions I get is "How can I get involved with helping to save our planet?" Well, in this book I share 100 actionable steps every reader can take to help the planet. Any of these steps would be a great place for you to start!

This book can be used as a resource and a tool to find ways you can personally play your part in living a more sustainable life that will benefit our world while learning how to pressure your government or local authorities to take action to help the planet. And it is also a book that will remind you during those darker moments that there is still plenty of good going on out there.

I've organized the book into chapters that I felt connected most to the key issues our planet is facing. There is some overlap between chapters—it's surprisingly difficult to split so many important topics into chapters—but I hope you'll find it flows splendidly and reveals how interconnected everything we do is.

IF THIS BOOK CAN ACCOMPLISH JUST TWO THINGS, I'D LOVE THEM TO BE THE FOLLOWING:

TO INSPIRE YOU WITH WAYS TO TAKE YOUR ACTIONS TO SAVE THE PLANET EVEN FURTHER, AND TO PUT A SMILE ON YOUR FACE AND GIVE YOU HOPE FOR THE FUTURE.

OCEANS

Since I grew up in Birmingham, which is one of the furthest places away from any coastline in the United Kingdom, the ocean seemed like a mythical land that was only present in stories or in books. When I finally got to visit it, the ocean became a place where many of my fondest memories were made: from searching for small fish and crabs in rock pools, to smashing into waves with my family.

Over the years, the more time I spent near the ocean, the more I learned about it and respected it. This also made me feel more compelled to help protect it. So why is our ocean so important? The ocean is home for up to 94% of all Earth's wildlife, covers over 70% of our planet's surface, and more than 80% of it remains unexplored.

The ocean absorbs approximately one third of the CO_2 we emit and has taken in over a nuclear bomb's worth of heat every single second for the past 150 years.

The ocean produces 50% to 80% of the oxygen on Earth and absorbs four times the amount of CO_2 as the Amazon rainforest, which is mostly thanks to marine plants that can store up to twenty times more carbon per acre than forests are capable of. The ocean is a carbon sink, meaning it stores vast amounts of CO_2, with studies calculating that 93% of all CO_2 is in it.

The ocean is full of weird and wonderful creatures. From the largest animal on our planet, the blue whale—whose enormous heart is the same size as a small car—to lobsters that communicate, interestingly enough, by peeing on one another.

The ocean is a life-support system for the planet. It provides millions of jobs and services across the world and helps regulate our climate. Unfortunately, our oceans are facing the harmful effects of plastic pollution, destructive fishing methods, and coastal developments, which are just a handful of the most aggressive threats in play. Thus, it's important we do all we can to protect our oceans and the precious ecosystems they are home to.

We'll explore conservation wins thanks to the hard work of ocean conservationists around the globe. We'll investigate some of the threats our oceans are facing and how activists are tirelessly fighting back against them. And we'll look into what you can do as an individual to play a role in helping to protect our oceans!

CHAPTER 1

Population Wins & Conservation Successes

Countries across the world are taking a stand for our oceans, helping to restore precious ecosystems by fighting against destructive industries. Conservation efforts in recent years have brought about wins that will help protect our oceans.

Since the ocean is so big, it can sometimes be difficult to regulate those who seek to exploit it. Luckily, there are good people out there who are ensuring the bad apples are held accountable for exploiting marine life and destroying their natural habitats by tirelessly working to help safeguard the sea life that thrives in our oceans. I want to introduce you to some of the activists and organizations protecting our oceans, good news about governments that are finally starting to take action, and wins for some of our beautiful marine species.

PROTECTING SHARKS

Hawaii has become the first state in the United States to ban shark fishing!

Contrary to how Hollywood movies make us feel about sharks, we should actually be excited that more sharks are being protected in our oceans. As a keystone species, sharks play an incredibly important role in our oceans, and are vital to maintaining the balance of marine life. Without them, ecosystems would fall apart as prey species overpopulate, disrupting entire food chains, and impacting oceans on a wider scale from habitat destruction to climate change.

As shark activist Paul de Gelder says, "People should not be afraid of having sharks in the ocean; they should be afraid of not having sharks in the ocean."

Our fear of them is largely irrational, too. In 2021, sharks killed just 9 people globally. By comparison, 11,000 to 30,000 sharks are killed every hour from industrial fishing.

From 2022 onward, shark fishing in Hawaii is banned, making it illegal for anyone to knowingly capture, entangle, or kill any species of shark in Hawaiian waters.

As reported by KITV4, violators will be charged with a misdemeanor and could be fined $500 for a first offense and up to $10,000 for repeat offenses. That fine applies to each individual shark captured or killed. Offenders could also have their commercial marine license, vessel, and fishing equipment seized or suspended. Let's hope more states in the United States follow Hawaii's precedent soon!

The United Kingdom has banned all shark fin products!

Shark finning is the act of removing fins from sharks and throwing the rest of the shark back into the ocean, mainly due to the demand for shark fin soup. The practice of shark finning has already been banned in United Kingdom waters, but trade has been allowed to continue under European Union law.

The United Kingdom became the first European nation to take this step with Animal Welfare Minister Lord Goldsmith hoping the ban would help boost shark numbers and send a message to the rest of the world that "we do not support an industry that is forcing many species to the brink of extinction." Hopefully, more countries soon follow in the United Kingdom's footsteps to help protect wild shark populations!

MARINE WILDLIFE CONSERVATION

Humpback whales have been thriving in Alaska with less traffic from cruise ships!

Something magical happened in 2020 that hadn't happened in more than a hundred years. Marine traffic declined so much in Glacier Bay, Alaska that researchers were finally able to study whale behavior in a quiet environment.

They observed humpback whales taking naps and singing a wider variety of whale songs. Since whales were able to hear one another over larger distances, mothers were able to leave their calves to play while they swam around in other areas to search for food. Less marine traffic means happier marine life!

Humpback whales are no longer considered endangered in Australia!

After strict global protections were put in place on commercial whaling, the humpback whale population has recovered to the point where they were removed from Australia's threatened species list in 2022.

A huge safe haven was created to help protect dugongs and other marine life in the northern Great Barrier Reef!

The World Wildlife Fund for Nature-Australia bought and shelved a commercial fishing license covering 38,610sq mi (100,00sq km) of the Great Barrier Reef and made it into a net-free safe haven for dugongs, dolphins, and many other members of this marine ecosystem.

This license would have typically been used by a commercial fisher to fish in the region for sharks, barramundi, and fingermark bream.

"It's not normal practice for a conservation organization to buy and shelve commercial net-fishing licenses. But it was a practical way to remove the threat of gillnets from a section of the reef incredibly important for threatened species," said Richard Leck, WWF-Australia Head of Oceans. Now, instead of allowing gillnet fishing to threaten the endangered species in the region, the WWF of Australia has created one of the largest safe havens for dugongs in the world!

STOPPING ILLEGAL FISHING

The tiny pacific island of Niue has declared a plan to protect 100% of its ocean from illegal fishing!

The ocean surrounding Niue is home to an abundance of precious marine life. Humpback whales, for example, migrate to Niue from Antarctica to give birth and to raise their young. However, they are threatened by illegal fishing, a huge problem in the Pacific Ocean.

Illegal fishing deprives some of the world's poorest coastal communities of crucial nutrition and income, causes global financial losses valued at up to $36.4 billion per year, and threatens endangered species.

To combat illegal fishing, Niue created a marine park the size of Vietnam around the island, which has strict laws against fishing illegally. If a vessel is caught, they can be heavily fined.

Niue Premier Dalton Tagelagi says he wants to remind people there is no other option, "We are doing our part to protect what we can for our future generation, just as our forefathers did for us."

Ecuador is using cutting-edge technology to fight illegal fishing around one of the world's most biodiverse places!

Ships that illegally fish often shut off the ability to track them and "go dark" so they can go unnoticed on radars. However, activists in Ecuador have found a way to track them through radio waves emitted by satellite phones or onboard navigation systems.

They utilize underwater monitoring systems that can identify a ship's location and the type of fishing gear it's using based on the sound it makes while sailing. As a result, in 2021, the Ecuadorian forces were able to catch a bunch of semi-industrial boats that were fishing illegally in waters reserved for small fishermen!

In Canada, a husband and wife created autonomous robots that can hunt down poachers!

Julie and Colin Angus set up Canada's first autonomous boat company, Open Ocean Robotics, to address the national and global challenge of monitoring and protecting oceans affordably, efficiently, and safely.

The boats are solar powered and provide real-time information to protect our oceans, safeguard Marine Protected Areas (MPAs), and facilitate the capture of illegally operating vessels

Not only can the robots patrol the ocean, they can also help detect and clean up oil spills, reduce greenhouse gas emissions, and improve the efficiency of ships out at sea!

RESTORING OYSTER REEFS

Billion Oyster Project is restoring New York Harbor's once flourishing oyster reefs!

This organization is on a mission to restore one billion oysters to New York Harbor by 2035. They're helping to bring biodiversity back while also educating the next generation of environmentalists at the same time.

New York was once home to over 200,000 acres (81,000 hectares) of oyster reefs, but due to ocean pollution and overharvesting, these reefs were wiped out in less than a hundred years.

Since launching in 2014, the Billion Oyster Project has restored 100 million oysters across 12 acres (5 hectares), established 15 reef sites where thousands of members of the community have helped build and monitor oyster reefs, and have diverted 2 million pounds (900,000 kilograms) of used shells from landfills, thanks to dozens of New York City restaurant partners.

Why are oyster reefs so important? Here are three reasons:

- *They help prevent flooding.*
  Oyster reefs can soften the blow of large waves and protect cities from storm damage since they act as natural barriers.
- *They clean the harbor.*
  One adult oyster can filter about 50 gallons (189 liters) of water a day, so imagine what one billion oysters could do!
- *They support marine ecosystems.*
  Oysters are ecosystem engineers that promote greater biodiversity and provide habitats for hundreds of different species.

STEPS YOU CAN TAKE

visit sambentley.co.uk/gnpe/oceans to access relevant links

O When planning a vacation, try to find one where you can either volunteer your time to support ocean conservation efforts or help support coastal communities. Think twice before booking a cruise, which can disrupt wildlife and pollute the ocean.

O Check out Billion Oyster Project to learn more about the importance of oyster reefs. Become a member, volunteer your time, or make a donation to help make New York Harbor thrive with biodiversity again.

O Follow shark activist Paul de Gelder on Instagram (@pauldegelder). Paul lost two limbs in a shark attack in 2009 and now advocates for protecting sharks. He'll give you helpful tips on what you can do to help protect our oceans' shark population, too.

These illustrations depict how adult oysters can filter water and improve its quality.

CHAPTER 2

Bottom Trawling Prevention

There's something awful happening at the bottom of our oceans that doesn't get spoken about enough. It's called *bottom trawling*.

The practice of bottom trawling involves dragging large, heavy nets along the sea floor to catch marine life. The nets are as wide as soccer fields and demolish vital habitats, releasing a huge amount of carbon as they disturb enormous swathes of ocean sediment. Over the course of 2019, bottom trawling is estimated to have released between 600 and 1,600 tons of carbon. Meanwhile, the combined amount of CO_2 released by global commercial aviation was estimated to be at 1,000 tons during the same time frame.

While approximately 25 million acres (10.1 million hectares) of forest are lost every year, a huge amount, more than 3.9 billion acres (1.6 billion hectares) of sea floor are lost due to trawling, with some of the largest trawlers catching up to 400 tons of fish per day. Thankfully, more awareness is being brought to this destructive fishing practice.

STOPPING BOTTOM TRAWLING IN ITS TRACKS

Greenpeace has created a barrier of boulders to stop trawlers from exploiting Marine Protected Areas off the coast of the United Kingdom!

After Greenpeace built a protective boulder barrier to obstruct destructive fishing practices such as bottom trawling in UK coastal conservation areas, the UK government announced that bottom trawling would be banned from four MPAs.

Within sensitive areas, the four bylaws prohibit bottom-towed gear like dredges, demersal seines, semipelagic trawls, and bottom trawls. These initiatives are being introduced using new powers under the Fisheries Act, the United Kingdom's first major domestic fisheries legislation in nearly 40 years.

This ban means that an abundance of marine life, including everything from crabs to flatfish, porpoises, and puffins, will be able to exist in a safer environment and contribute to healthier ocean ecosystems.

There is still work to be done to help fully protect the MPAs from other destructive industrial fishing vessels, but wins like this are hard to come by and worth celebrating and sharing! They're also proof that taking action, protesting, and making noise about these important issues can help create huge, positive changes in our oceans.

CURBING UNREGULATED FISHING

Sea Shepherd is working around the world to protect our oceans' endangered marine life!

An international nonprofit organization called *Sea Shepherd Conservation Society* is on a mission to protect the world's oceans from illegal exploitation and environmental destruction.

Started back in 1977 as a grassroots campaign, today the movement spans more than 20 countries and works with law enforcement agencies and governments around the world to bring poachers to justice and end illegal, unreported, and unregulated fishing.

But it's Sea Shepherd's direct action that makes it different from other organizations working in the field. Far beyond petitions and marches, Sea

Shepherd's innovative direct-action techniques have seen crews (staffed mostly by volunteers) stand up to seal hunters, haul in miles of illegal fishing gear, and rescue more than 6,000 whales from Japanese harpoon ships in dangerous Antarctic waters.

One of their biggest battles is against bottom trawling and destructive fishing, often assisting nations unable to defend their own coastal waters against illegal practices due to a lack of resources. To do this, their crews track, record, and even chase down ships caught working outside the law and bring them to justice.

To date, Sea Shepherd has been part of more than a hundred direct-action campaigns aimed at protecting ocean species and combating illegal fishing.

REMOVING HARMFUL DEBRIS FROM OUR SEAS

The Ocean Defenders Alliance (ODA) is cleaning up debris in the Pacific Ocean that's detrimental to sea life!

ODA has been actively working with communities and volunteers to remove harmful debris from the sea since 2002.

Today, this nonprofit organization, alongside teams of volunteers of all ages and from all walks of life, works to tackle things like abandoned fishing nets, traps, lines, plastic, and other man-made debris threatening ocean wildlife and habitats.

Mainly based around California and Hawaii, it's funded entirely by donations from those who are concerned about the problem of marine debris and want to be part of the solution. From technical divers to those who can't actually swim, ODA brings communities together to work toward a common purpose: debris-free seas.

Discarded fishing gear makes up a large amount of the debris collected by ODA, which is why the team also runs education and outreach programs—not just with schools, but also with fishermen, restaurants, and seafood communities—to raise awareness about the dangers of destructive fishing.

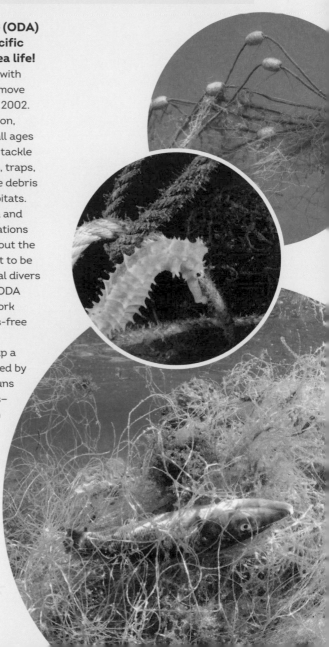

STEPS YOU CAN TAKE

visit sambentley.co.uk/gnpe/oceans to access relevant links

○ If you're able to, shifting toward a plant-based diet is an incredible way to reduce our ecological impact and help our oceans recover. If you still want to eat fish, consider buying from local fishermen, as they aren't using bottom trawling to capture fish.

○ Discover more about bottom trawling in documentaries like *Seaspiracy* on Netflix to become more aware of what's happening out at sea and what we can do as individuals to help prevent it.

○ Push for a global ocean treaty by encouraging your local government to enforce no-catch marine reserves that would allow threatened species and sensitive ecosystems to be protected, with the goal of protecting 30% of our oceans by 2030.

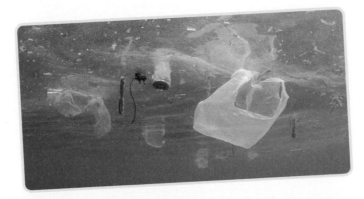

CHAPTER 3

Ocean-Cleanup Innovations

The fact that there's a crazy amount of plastic in our oceans likely won't surprise you. You've probably heard stories of plastic turning up even in the most remote areas of our world, far away from human civilization. This is no surprise when the equivalent of a garbage truck of plastic is dumped into our oceans every minute.

This litter is a result of intentional and accidental human behavior, with land-based activities contributing to roughly 80% of marine pollution. The careless disposal of waste—whether that's waste released from dump sites near the coast or the common littering of beaches—are just a few examples of how our oceans are polluted by humans.

Even litter dropped far away from the coast, like bottles, plastic wrap, and cigarette butts, can make their way to the ocean through drains, rivers, and other waterways, as do materials flushed down toilets.

Of the pollution that comes from sea-based sources, 20% is due to abandoned, lost, or discarded fishing gear. For example, a 2018 study found that fishing nets accounted for 46% of debris found in the Great Pacific Garbage Patch.

But the good news is the fight against plastic pollution is on, and it's only getting stronger every year. Out at sea and on land, organizations are getting creative with collecting, recycling, and combating plastic waste at the source.

GETTING PLASTICS OUT OF OUR OCEANS

Seabin Project has created an amazing invention that helps remove litter from our oceans!

A surfer in Australia invented a floating trash can that catches up to 50,000 plastic bottles a year! And not only plastic bottles, but oil, detergent, and paper, too! (Essentially, anything floating in the water that's not supposed to be there!)

This magical device is called a *Seabin*, and it was created by the team at the Seabin Project to help tackle plastic pollution that's plaguing our marine environments. It's perfect in controlled environments like marinas, ports, and yacht clubs where no ocean storms are present.

Since July 2020 a fleet of 16 Seabin units have helped capture more than 100 tons of marine litter in Sydney alone. And on a global scale, they help collect 4.2 tons of marine litter per day across 53 countries!

For those worrying about sea life potentially getting trapped in these devices, have no fear, as the team at the Seabin Project also monitors each Seabin to ensure it's not causing damage to the populations of local marine life.

The organization's goal is to build more Seabins using plastic they recover from the ocean, which will make it an even more environmentally friendly invention. They're also working on scaling up the Seabins to be about 10 times the size of the current model's design, so they can help collect more plastic waste.

The ultimate goal here is to create a world where Seabins eventually won't be needed, as we strive to help create pollution-free oceans for future generations!

More than 220,500lb (100,000kg) of plastic has been removed from the Great Pacific Garbage Patch!

The Great Pacific Garbage Patch is the largest accumulation of ocean plastic in the world and is located in the Pacific Ocean between Hawaii and California. It takes up an area twice the size of Texas or three times the size of France!

The Ocean Cleanup, a nonprofit organization, is developing and scaling technologies to rid the world's oceans of plastic, especially in the Pacific. It's being called the largest cleanup in history.

One of their plastic-collecting innovations, System 002, has swept an area of ocean comparable to the size of Rhode Island, gathering tons of plastic from the ocean's surface.

Work has started on their new and expanded System 03, which is 3 times the size of System 002!

The Ocean Cleanup has also developed technology to intercept pollution before it reaches the ocean in the first place, such as the Interceptor, which can work autonomously and extract up to 110,000lb (50,000kg) of plastic per day from rivers.

They've also been experimenting with a waste-capturing Trashfence in Guatemala's Rio Motagua Basin, home to what is said to be the world's most polluted river. The Interceptor Trashfence is inspired by avalanche protection systems and is specialized for flash flood environments.

REMOVING OTHER LITTER FROM OUR OCEANS

Artists in Kenya are turning old, washed-up flip-flops into works of art!

Ocean Sole is a social enterprise that upcycles washed up flip-flops that are found along the beaches and waterways of Kenya.

This group collects flip-flops found from the ocean, gives them a thorough wash, and then presses them into blocks, which are carved into various sculptures. These sculptures are then smoothed and sanded down before details are added and the amazing artworks are finished.

Ocean Sole aims to recycle a million flip-flops per year, recycle over 1 ton of styrofoam a month, and save over 500 trees a year by using flip-flops instead of wood.

The group also positively impacts over a thousand Kenyans by employing them to collect the flip-flops, providing steady income to low-income families. On top of this, they give 10% to 15% of their revenue to beach cleanups, educational programs, and other ocean conservation groups.

STEPS YOU CAN TAKE

visit sambentley.co.uk/gnpe/oceans to access relevant links

O Eliminate items from your life that contribute to marine litter. Take a reusable water bottle out with you, so you don't need to purchase a plastic bottle. This can have a massive impact on preventing plastic bottles from entering our marine environments.

O Donate to the Seabin Foundation, an Australian charity set up by the Seabin Project that you can find out more about on their website.

O Follow the Ocean Cleanup on social media (@theoceancleanup) and learn more about how they're helping rid our oceans of plastic.

O Roughly 80% of marine pollution originates on land, so get involved locally with addressing land-based pollution. Explore Marine Litter Solutions' website to learn how to help address marine litter, or visit the UNEP website to become a part of the generation that helps heal our planet.

O Support the Kenyan artists at Ocean Sole by visiting their website and buying one of their awesome flip-flop sculptures!

O Attend a local beach cleanup if you live near a coast. Preventing plastic from getting into local rivers is just as valuable, too! See if there are any opportunities to get involved in a cleanup in your local area.

LAND

Land accounts for under 30% of the Earth's surface, but it's home to 100% of the human population, and carries out vital processes that keep our ecosystems and atmosphere functioning properly. We rely on land for our food, our homes, and our way of life.

A critical part of our land is our soil. Soil is one of the most biodiverse ecosystems on the planet that contains and nurtures trillions of living organisms. Around 40% of all life on Earth depends on soil. Approximately 95% of the food we eat comes from soil, and together with plants, it also absorbs around a third of our carbon emissions.

All this carbon being locked up in the soil means it doesn't enter the atmosphere as CO_2, and doesn't further exacerbate global warming. That's why it's so important to educate ourselves and protect this part of our landscape, which might appear unproductive to the untrained eye, but are vital to our survival.

Landscapes, like peatland, act as huge carbon stores, too. Scottish peatlands alone store 1.7 billion tons of carbon, which is the equivalent of 140 years' worth of Scottish annual greenhouse gas emissions.

As the human population continues to grow, the need for more area to grow food and build new homes also grows. This can often come at a huge cost to the environment, and biodiversity on land is threatened as ecosystems are altered or destroyed. Species are put at risk of extinction as they're driven from their habitats. Biodiversity brings balance, which helps support the future of food security and water supply, while also allowing natural cycles to function as they should.

One such practice that's helping in the fight is agroforestry, which is when trees are planted among agricultural crops. This allows for more diversity on agricultural land, which promotes and protects biodiversity both above and below ground. It also stabilizes the soil, helping minimize soil erosion and conserving organic matter.

If it is allowed to, US soil can capture 275 million tons of carbon dioxide–equivalent greenhouse gasses every year. However, when the chemical processes within the soil aren't allowed to function properly and desertification or degradation occurs, this cannot happen, and more carbon is left in the atmosphere as CO_2, which we all know is not a friendly gas in excess amounts.

Over this chapter, we'll look at the many threats our land is facing, from deforestation and desertification to plastic pollution. We'll explore the many people and organizations who are fighting to protect and become stewards of our land and create a more equitable and regenerative future.

CHAPTER 4
Deforestation & Desertification

Deforestation

While you may be familiar with the definition of deforestation as the clearing of sections of trees, you may not be aware of the ripple effects this activity can have on the surrounding environments where it occurs.

The most notable example of extreme deforestation is in the Amazon rainforest. The Amazon is being deforested at a rate of one soccer field every 12 seconds. With roughly half the world's species calling the Amazon rainforest home, these species are being put at risk every day.

Orangutan populations in Borneo rainforests have decreased by over 50% in the past 60 years due to the deforestation of their habitats. This is driven by the palm oil industry—the most consumed vegetable oil on the planet—which is causing hugely detrimental effects on global biodiversity as it threatens these orangutans, who act as ecosystem engineers that modify and maintain their habitats.

It's estimated that if the rainforest were left to grow naturally, it would provide more economic value through its harvestable products, such as nuts and fruits, than the area produces when cleared for cattle and agriculture.

On top of the potential economic benefits of protecting these rainforests, their protection would also ensure that global biodiversity is preserved. Even more than the desire to maintain the aesthetic value of these woodlands, high biodiversity is associated with environmental stability and, thus, resource security for the future.

Desertification

Perhaps a lesser known consequence of climate change is desertification, which is the degradation of fertile land into infertile desert soil. Over 75% of land is already degraded in some way and this could increase to over 90% by 2050. This puts global food security at risk, as it decreases the amount of arable land available to our global agricultural industries.

Desertification is primarily caused by human activities, like the mismanagement of agricultural land, overgrazing, and deforestation. These activities strip the protective vegetation—which are the plants that allow water to infiltrate the soil effectively—and prevent the maintenance of healthy soil structures.

According to recent predictions, there could be 1.2 billion climate refugees by 2050, mainly due to desertification. As the Earth warms and there is less precipitation, desertification will degrade more and more landscapes. As these landscapes become uninhabitable because agricultural livelihoods are made impossible or temperatures rise to unlivable conditions, we will see a rising number of people displaced toward cities as they search for better conditions and work opportunities.

One of the most direct ways to combat desertification is for farmers to use regenerative agricultural practices. These practices work to preserve the quality of the soil, while still producing high enough crop yields to support our population. These methods can be as simple as no-till farming, crop rotation, or incorporating smart farming technologies.

In this chapter, we'll look at some positive ways in which experts around the world are working to combat desertification and deforestation to keep our planet and our people safe and happy.

PLANTING TREES AROUND THE WORLD

Around the world, huge tree planting projects are helping revitalize millions of acres of forests!

In July 2022, the United States government announced they would plant more than one billion trees over the next decade to address a reforestation backlog of four million acres across its national forests.

In India, during the start of the pandemic, two million people gathered to plant 20 million trees in a single day as part of India's pledge to increase its forest coverage to 235 million acres by 2030.

Thanks to a $1.35 billion tree-planting project that was launched by the Ethiopian government, over 350 million trees have been planted in Ethiopia in just over 12 hours!

At the time of writing this, the current Guinness World Record for the most trees planted by one person in 24 hours is attributed to Antoine Moses, who planted 23,060 trees in one day in Alberta, Canada in 2021.

SEEDBALLS

Communities in Kenya are creating seedballs to bring millions of acres of degraded land back to life!

Local organization Seedballs Kenya is behind this awesome initiative and have created small balls that contain a seed that can grow to become a tree.

How do seedballs work? Waste charcoal dust is collected and the seed of a native tree is mixed in with that dust. When the seedball is thrown, this carbon coat protects the seed from hungry animals until rain arrives. An abundance of nutrients that is present in the charcoal dust helps give the seed a good start.

Kenya is losing thousands of local trees every year that people rely on for fuel, and traditional reforestation projects take time and are relatively costly. By using seedballs, tons of trees can be planted at once, costing as little as $20 per acre to restore Kenyan dryland.

Even if only a handful of the seeds grow into trees, the effort is worth it.

Seedballs Kenya is striving to make the process as accessible as possible. They've even created slingshot competitions so young people can help distribute the seedballs and make the tree planting process more fun.

So far, over 26 million seedballs have been distributed in Kenya, and when trees reach maturity within 4 to 14 years, they can provide the community with sustainable forest products, including charcoal and building materials, and help mitigate climate change impacts and improve water availability.

SEED-FIRING DRONE

Did you know there's a drone that can fire seed pods and plant up to 40,000 trees every day to help combat deforestation?

In the last 30 years, the world has lost over 1 billion acres of trees. This compelled a start-up in Australia called AirSeed Technologies to develop technology that can plant trees with a drone. The CEO claims it's 25 times faster and 80% cheaper than traditional methods.

Each drone is loaded with seed pods that are customized to each local habitat. It can plant two seed pods every second, which works out to 40,000 trees a day with a team of two people.

The location of every seed pod that is dropped is able to be tracked by AirSeed Technologies' software, allowing them to track the progress and success of each seed pod.

Every seed pod is full of nutrients and contains a carbon coating that helps absorb rainwater and protects the seed from birds and rodents. This gives the seed the best chance possible to successfully grow into a tree!

AirSeed Technologies has a goal of planting 100 million trees by 2024 to help mitigate the level of warming over the next 30 years. Their goal is to give the following generations a fighting chance of having a sustainable environment to inherit, along with a greener planet too!

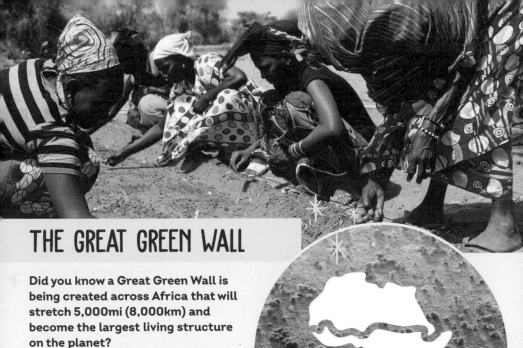

THE GREAT GREEN WALL

Did you know a Great Green Wall is being created across Africa that will stretch 5,000mi (8,000km) and become the largest living structure on the planet?

The African-led Great Green Wall initiative was launched in 2007 and aims to restore Africa's degraded landscapes and transform millions of lives for the better.

The ambitious project is working to restore 247 million acres (100 million hectares) of degraded land, sequester 250 million tons of carbon, and create 10 million green jobs.

This enormous undertaking will help communities living along the wall by:
- *Creating more fertile land, a precious natural asset.*
- *Providing economic opportunities for the region's young people.*
- *Improving food security for the millions of Africans who go hungry every day.*
- *Developing climate resilience in a region where temperatures are rising fast.*
- *Creating a new world wonder spanning 5,000mi (8,000km).*

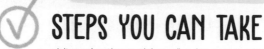

STEPS YOU CAN TAKE

visit sambentley.co.uk/gnpe/land to access relevant links

○ Sign petitions that support the mandatory labeling of all products that contain ingredients like palm oil, which are produced as a direct result of deforestation. Holding large corporations accountable for their participation in these destructive activities is the first step in initiating change.

○ Encourage your government to protect important local woodlands from real estate developers. Alternatively, sign petitions to designate specific protected areas anywhere new home expansion is important to help meet housing requirements for our growing population.

○ Volunteer at tree planting organizations in your local area and abroad. Counteracting deforestation comes in two parts: petitioning against mass deforestation, while also engaging in the replanting of trees whenever possible. When the right species are planted in the right locations, trees can help absorb masses of CO_2 from the atmosphere, which helps minimize the effects of climate change made worse by these deforesting activities.

○ If you'd rather leave the work to the experts, you can donate to tree planting organizations like the Eden Reforestation Projects that works with local communities to restore forests on a massive scale. Be sure to visit their website. You can also help out by donating to WeForest. This organization has programs in place where you can sponsor a forest as a company or as an individual, plant trees locally, or make a donation to help combat deforestation.

○ You can support Seedballs Kenya and help regreen Kenya by visiting the donation page on their website, and by following them on Instagram (@seedballskenya).

CHAPTER 5

Regeneration & Protection

The health of a soil is dictated by its structure. Ideally, soil should have a healthy balance of organic matter, inorganic elements, and pore space to enable it to drain water and supply nutrients effectively enough to grow the plants that our ecosystems rely on. However, when soil is degraded by poor farming practices and other human activities, this ideal soil structure is disrupted and the soil cannot function as it should.

Regenerative agriculture is the idea of restoring soil's organic matter and rebuilding soil biodiversity. This can increase carbon drawdown into the soil, improve water circulation, and bring about better crop yields and overall crop quality.

The foundation of most of these regenerative practices stem from agroecology. This is the idea of basing agricultural practices around ecology instead of around mass production goals, which requires taking the unique ecology of each location into account when deciding how to treat the soil, which plants to grow, and how often to grow them.

Recent studies have shown that farming practices that incorporate regenerative methods can be up to 78% more profitable than those that do not. Regenerative agriculture means merging farming and natural resource conservation profitably. We could see a boost in agricultural economies as well as happier soil and environments when these methods are used. Speaking of happier soil and environments, I have some amazing things to share with you in this chapter!

KENYAN BUNDS

Semicircle holes in Kenya are helping make degraded landscapes green again, while also cooling down the planet at the same time!

These moon-shaped holes are called *bunds*, and Kenyan farmers have dug over 200,000 of them to save their land from drought.

What do bunds do? Bunds capture rainwater that would otherwise get washed away over the dry soil.

This rainwater can then seep into the soil, allowing vegetation to grow in and around the holes, meaning large areas can be brought back to life in a short amount of time. Bunds also help create more rain—as the vegetation that grows creates evapotranspiration, which leads to clouds forming that often release precipitation—and prevent soil erosion.

By the end of this century, more than two billion people are projected to be exposed to increased droughts, so solutions like this are super important in the fight against water and food scarcity. By regreening dry land, we can cool the earth and provide water, food, biodiversity, and a better life for millions of people and animals.

The organization behind this amazing work is called *Justdiggit*, and they're on a mission to regreen Africa over the next 10 years.

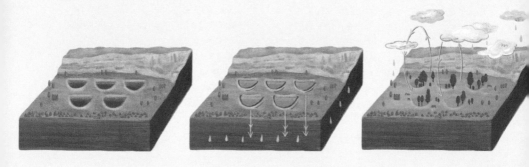

GOATS PREVENTING WILDFIRES

Just when you thought you couldn't love goats anymore, you find out they're able to help prevent wildfires!

It's true, goats are helping prevent wildfires in California! How are they doing this, you ask?

Well, every year, 400 or so goats help clear out long grass and weeds in hard to reach areas around West Sacramento in an effort to help stop the spread of wildfires.

This is a more environmentally friendly option, using goats that serve as natural mowers and whose mouths are able to get to all the overgrown spots across uneven terrain, rocks, and forested landscapes. It's also a lot quieter than traditional mowers, so it helps cut down on noise pollution, too.

Most wildfires in the United States are started by human activities, and overgrown forests and thick vegetation can fuel fires that grow out of control. With these goats playing such a big part in the city's wildfire suppression program, they're able to help prevent wildfires, protect the environment, and keep people safe.

LAND PROTECTION & RECOVERY

Around the world, funding is being poured into protecting nature, which is helping it recover!

In June 2022, the government of Colombia, along with World Wildlife Fund (WWF) and other partners, announced it was committing $245 million to protect 79 million acres (32 million hectares) of iconic Colombian landscapes and seascapes. Like many countries, Colombia has a target to protect 30% of its land and 30% of its seas by 2030, and this commitment to long-term conservation will certainly help them be well on their way to hitting that "30 by 30" target.

Protecting nature is a great way to let our planet heal, but we can also play a more active role in helping nature recover, too.

In England, five nature recovery projects were launched to tackle wildlife loss and improve public access to nature. These projects cover over 245,000 acres (99,000 hectares) of land across England and will improve the landscape's resilience to climate change, providing natural solutions to reduce carbon and manage flood risk.

The aim of these multipartnership projects is to restore habitats and

introduce corridors and stepping
stones to help wildlife populations
peacefully migrate between urban and
wild spaces.

And the power isn't just in the
government's hands when it comes
to protecting our precious habitats.
Thanks to an overwhelming number
of public donations, Rainforest
Trust was able to protect more
than 1 million acres of habitat in
the first 3 months of 2022 and is
on track to protect an additional
125 million acres (50 million
hectares) by 2025.

REGENERATIVE AGRICULTURE

Support is on the way to help the agricultural industry play a major role in transitioning to a nature-positive economy!

The farmers and people who work in the agricultural sector need as much support as possible to help them. So that's why it's great news that the Agri E Fund was launched in the United Kingdom by Virgin Money for farmers looking to make their businesses more sustainable to help create a greener future. The fund consists of $245 million (£200 million) of lending, which will support farmers adapting to environmental changes. It also happens to be the first UK green product specifically designed for UK farmers.

A landmark study by scientists from the UK Centre for Ecology & Hydrology even found that nature-friendly methods of farming, such as dedicating sections of farmland for nature recovery can significantly increase biodiversity without damaging food production.

STEPS YOU CAN TAKE

visit sambentley.co.uk/gnpe/land to access relevant links

O You can support Justdiggit in their mission to regreen Africa by buying a bund for around $9 (£7) to help out! Visit their website to donate. By buying bunds you can help invest in a cooler climate and directly support an African farmer who digs the bunds.

O Support farmers and organizations who are promoting and using regenerative practices. Check for the Land to Market verified seal on products at the supermarket. These are products that have been grown on land that's actively regenerating.

O Plant your own garden at home if you can. Getting your vegetables from your own garden is the most sustainable thing you can do, and it will help regenerate the microhabitat of your backyard.

CHAPTER 6
Plastic Pollution

The negative impact plastic is having on our planet is probably the one thing you don't need to hear more about. Instead, I thought I'd talk about the logos associated with recycling and some of the myths surrounding them.

Green dot: Most people in Europe assume this means a product is recyclable, right? Sadly this is not the case. It simply means the producer of the product has made a financial contribution toward the recovery and recycling of packaging in Europe and not that the packaging is fully recyclable.

Universal recycling symbol: Surely this one means a product is recyclable? It might be. The recycling symbol indicates a package is capable of being recycled in areas where collection facilities for the material exist. So depending on where you live, it could be recyclable, or it might not be.

Have you ever noticed the number inside the recycling symbol? These numbers indicate what type of plastic the item is made from, not that it's recyclable. They're called *Resin Identification Codes* and they go from 1 to 7, defining the type of resin used. In the UK, the most accepted types of plastic at curbside pickups are numbers 1 and 2, but this may differ based on where you live.

The fight against plastic pollution is truly underway, from governments finally taking a firmer stance, to innovators across the world getting creative with ideas to cut down or capture plastic waste. That's what we'll take a look at in this chapter. Oh, and also a plastic-eating enzyme that could eliminate billions of tons of landfill waste!

BANNING SINGLE-USE PLASTIC

While individuals can help tackle plastic pollution, it's crucial for our governments to implement new policies to tackle it across the world!

Thankfully, cracking down on plastic pollution has been gathering more momentum and world leaders seem to finally be taking some action.

In July 2022, India imposed a ban on single-use plastic items, such as straws, cutlery, earbuds, packaging films, plastic sticks for candy, and cigarette packets, among other products.

California has also signed a law that contains the most sweeping restrictions on single-use plastics and packaging in the United States. The state has taken nation-leading steps to cut plastic pollution and hold the plastics industry accountable for their waste.

The new legislation requires all packaging in the state to be recyclable or compostable by 2032, cutting plastic packaging by 25% in 10 years and requiring 65% of all single-use plastic packaging to be recycled in the same timeframe.

Additionally, the legislation shifts the plastic pollution burden from consumers to the plastics industry by raising $5 billion from industry members over 10 years. This will assist efforts to cut plastic pollution and support disadvantaged communities hurt most by the damaging effects of plastic waste. The manufacturers will also have to pay for the recycling programs.

This new law will help eliminate 23 million tons of plastic over the next 10 years!

BEACH VACUUM

Engineers in Canada invented a beach vacuum that sucks up small bits of plastic on beaches!

It's called the *HO Micro*, invented by the team at Hoola One, and it sucks up a mixture of plastic and sand. This mixture is then separated by buoyancy within the machine, allowing the machines to return the sand to the beach and extract just the plastic.

The HO Micro can clean more than 37lb (17kg) of sand per minute, and collects plastic from 0.0004 to 3in (0.001 to 7.5cm) in size. It works on any type of soil, can collect deeply buried plastics, and even works on rugged terrain by helping to suck up those hard to reach bits of plastic.

Of the floating plastic that enters the ocean, studies estimate that 97% of it is redirected to shorelines and beaches, so solutions like this are really important to help restore coastal environments.

LASER-FRUIT MARKINGS

Are you sick of seeing those small plastic stickers on your fruits and vegetables from the grocery store?

Well, a company in Spain called *LaserFood* has created laser labeling technology that could eliminate them! And like the name of the technology suggests, a laser is used to label the produce instead!

It's called *natural branding*, and it uses a high-definition laser that removes a bit of pigment from the skin of the produce to create a label. It's a contactless method that is completely safe, and it doesn't have a negative impact on the taste, aroma, or shelf-life of the fruit or vegetable. It's even been approved by EU organic-certifier Skal.

As natural branding doesn't use plastics, glues, inks, dyes, or packaging, it creates less than 1% of the carbon emissions needed to produce a sticker of a similar size, and it can be applied to practically all fruits and vegetables, including mangoes, sweet potatoes, avocados, and much more!

TOPUP TRUCK

An entrepreneur in London converted an old electric milk truck into a zero-waste store on wheels to help fight plastic waste!

It's called the Topup Truck, and it brings products straight to the door of local communities without all the usual plastic packaging. It was created by Ella Shone as a way to help make ecofriendly living more accessible.

So how does the Topup Truck work? People simply book the truck to visit their house or check its delivery schedule and come outside to fill up their own containers with a wide variety of goods like grains, dried fruit, household products, and more.

So far, Ella's converted milk truck has saved over 44,000 pieces of plastic from being used, while also empowering local communities to become more sustainable.

UK and US citizens produce the most plastic waste per person, much of which is exported to poorer countries, polluting their lands and oceans. So local solutions like this are really important to help create a better future across the world!

PLASTIC-EATING ENZYME

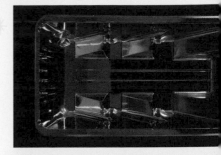

A plastic-eating enzyme was created at The University of Texas that could eliminate billions of tons of landfill waste!

Certain plastics take centuries to degrade, meaning that waste piles up in landfills and pollutes our natural lands and waterways. But this enzyme variant–created by engineers and scientists at the University of Texas–can break down environment–throttling plastics in just a matter of hours to days.

Essentially, the enzyme breaks down plastics into smaller parts and spits out recycled, new plastics, that are ready to be reused.

This means the recycling process could be supercharged on a massive scale, allowing major industries to reduce their environmental impact by recovering and reusing plastics at the molecular level.

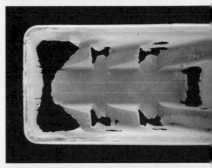

The result? The beginning of a truly circular plastics economy that would see the world tackle a material that currently makes up 12% of all global waste.

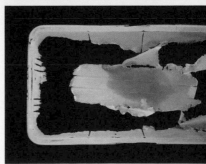

STEPS YOU CAN TAKE

visit sambentley.co.uk/gnpe/land to access relevant links

O If you're interested in shopping without plastic, you can find the locations of bulk stores on websites such as Zero Waste Home or Litterless, if you're in the United States. If you're outside of the US, you can search for *bulk stores near me* or *zero-waste shops near me* to find local solutions. Remember to bring your own reusable cloth bags, containers, or jars to the store!

O Shopping at farmers' markets is fantastic for many reasons, one of them being that you can buy local goods without all the plastic packaging or those pesky little stickers! Check out Local Harvest to find your local farmers market, if you're in the United States and Farming UK, if you're in the United Kingdom.

O Avoid buying new whenever possible. Usually, new products come in an abundance of plastic packaging and require a whole new set of materials to make them. By looking for what you need at local secondhand shops, or on websites such as Facebook Marketplace or Freecycle, you can avoid all that plastic and likely save a bunch of money, too! A personal favorite of mine is the Buy Nothing Project, the world's biggest gift economy, which you can learn more about at their website.

O If you're in the United Kingdom, you can use the website Recycle Now to find out what you can recycle at home and where the nearest drop-off site is for those items your council doesn't recycle curbside.

CHAPTER 7

Indigenous Wins

Over the last several centuries as European settlers colonized other lands and imposed their cultures onto the indigenous people already existing there, many traditional practices were lost. While colonizers saw the land as something to be used as they pleased, indigenous communities acted as stewards of the land and protected it in a way that took local biodiversity and ecology into account. These Indigenous Knowledge systems are recognized ideologies that were developed before modern scientific knowledge.

As the world has become more Western centric, large cities have taken over and indigenous peoples haven't had access to their native lands. As a result, many of their age-old practices have been lost in our modern society. Indigenous communities provide a wealth of knowledge on how to care for our planet by keeping our environment constantly in mind. While indigenous peoples account for 5% of the world's population, the UN estimates that the land indigenous peoples live on is home to 80% of the world's remaining biodiversity.

Policymakers have finally started consulting indigenous people and incorporating their knowledge into the systems where modern knowledge has failed. There have been motions in recent years to return stolen indigenous lands, which a large portion of colonized countries such as Canada, the United States, and Australia currently occupy.

While indigenous knowledge is increasingly recognized, it is still rarely being consistently taken into account by policymakers and large corporations. Returning this land to indigenous communities is not only doing what's right, it's also benefitting humanity by opening the door to a whole new set of knowledge that can be implemented to combat climate change.

RETURNING LAND

Following centuries of land being stolen from Native Americans, the United States has started returning ancestral Native American lands to their indigenous owners!

In 2021, an 18,000-acre (7,280-hectare) bison range in northwest Montana was transferred from the US Fish and Wildlife Service back to the Confederated Salish and Kootenai Tribes of the Flathead Reservation in an important move to restore tribal homelands.

The land was unlawfully taken from the tribes in 1908 by the United States. It was turned into the National Bison Range to protect the American bison from extinction, following the desecration of

the species that was driven by American settlers and systematic hunting.

In April 2022, 465 acres (188 hectares) of tribal land in Virginia, which was seized by European settlers in the seventeenth century, was returned to the Rappahannock Tribe in what was called "a historic victory for conservation and racial justice."

According to the Department of the Interior, the tribe plans to create trails and a replica sixteenth-century village to educate visitors about Rappahannock history and conservation efforts, as well as train tribal youth in traditional river knowledge.

Restoring land ownership to indigenous communities isn't just

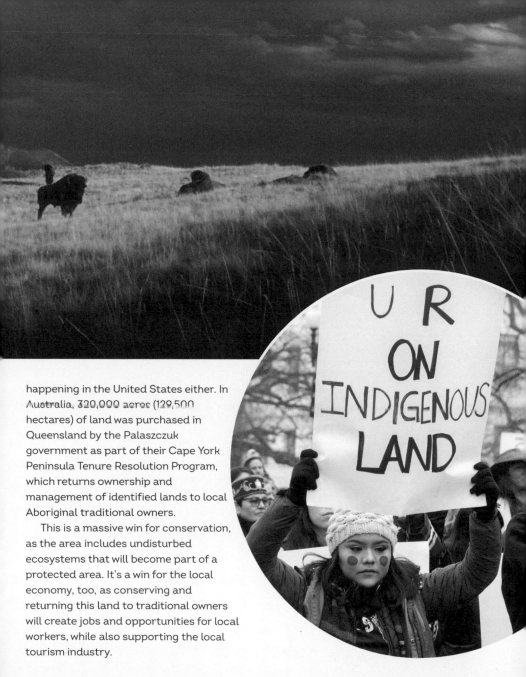

happening in the United States either. In Australia, 320,000 acres (129,500 hectares) of land was purchased in Queensland by the Palaszczuk government as part of their Cape York Peninsula Tenure Resolution Program, which returns ownership and management of identified lands to local Aboriginal traditional owners.

This is a massive win for conservation, as the area includes undisturbed ecosystems that will become part of a protected area. It's a win for the local economy, too, as conserving and returning this land to traditional owners will create jobs and opportunities for local workers, while also supporting the local tourism industry.

CO-MANAGING BEARS EARS

Native tribes are finally being allowed to help manage Bears Ears National Monument!

In recognition of tribal knowledge about their ancestral lands, the US government reached a historic agreement to give five Native American tribes more input in the management of Bears Ears National Monument in Utah.

The cooperative agreement was made between the Bureau of Land Management, US Forest Service, and the Hopi Tribe, the Navajo Nation, the Ute Mountain Ute Tribe, the Ute Indian Tribe of the Uintah and Ouray Reservation, and the Pueblo of Zuni.

The agreement stated "the Tribal Nations, and particularly Tribal Elders, have important knowledge, local expertise, and an understanding of the spiritual significance of the Bears Ears region beyond the physical environment."

Carleton Bowekaty, co-chair of the Bears Ears Commission and lieutenant governor of the Pueblo of Zuni, said in a statement, "Instead of being removed from a landscape to make way for a public park, we are being invited back to our ancestral homelands to help repair them and plan for a resilient future."

This is a major step in the right direction, and hopefully we'll be seeing more governments involving native tribes in managing their ancestral lands soon!

STEPS YOU CAN TAKE

visit sambentley.co.uk/gnpe/land to access relevant links

O Support farmers and organizations that are using indigenous practices by choosing their products over companies using less sustainable farming practices.

O Consider buying local indigenous products like artwork, jewelry, and food, before considering other alternatives.

O Listen to what the indigenous communities have been telling us, not only their knowledge, which can help us, but also their lived experiences, in order for us to better understand where they are coming from. Healing begins with acknowledgement, so listening to understand and recognize these injustices is a good place to start.

O Sign petitions to support indigenous peoples having their lands returned to them. Organizations like LANDBACK are doing great work to raise awareness and help return land to indigenous communities. Returning land to its rightful owners will allow it to be cared for properly and allow for justice to be delivered, as we return what was taken so long ago.

FOOD

With the global population expected to hit almost 10 billion by 2050, ensuring our global food system can cope is absolutely vital. Right now, though, the current setup is having a detrimental impact on the planet.

First, let's talk about emissions. From production to consumption, it's estimated the global food system contributes to 34% of all man-made greenhouse gas emissions. With food generating an average of 2 tons of carbon dioxide equivalent (CO_2e) emissions per person annually, that's more than all emissions from transportation, heating, lighting, and air conditioning combined.

Emissions aren't the only issue, though; the current way we grow and process food is incredibly unsustainable, resulting in significant resource depletion (such as destroying thousands of acres of rainforest to make way for agricultural land) and increased loss of biodiversity.

So what can we do about it? Many scientists and researchers have already outlined a new vision for a sustainable global food system of the future—and it largely includes taking a long, hard look at our diets. According to them, reducing meat and dairy consumption is the single biggest way individuals can lessen their impact on the planet.

The meat-and-dairy industry is a major source of greenhouse gas emissions. Livestock production accounts for around 14.5% of all human-induced emissions, making it one of the biggest contributors to climate change.

Dr. Marco Springmann of the Oxford Martin Program on the Future of Food, said, "What we eat greatly influences our personal health and the global environment. Imbalanced diets, such as diets low in fruits and vegetables and high in red and processed meat, are responsible for the greatest health burden globally and in most regions. At the same time, the food system is also responsible for more than a quarter of all greenhouse gas emissions and therefore, is a major driver of climate change."

And it doesn't stop there. We need to look at our relationship with food waste, a common issue across all stages of the food chain, as well as learn how to make the most of regenerative agriculture to replace what's been lost.

If we are armed with the right knowledge and a willingness to take action, we stand the best chance to feed the planet and save it at the same time.

CHAPTER 8

Food Waste

Food waste is a growing problem around the world, and it's time to take action. From farm to table, around 30% of food is lost or wasted along the chain—enough sustenance to feed three billion people. And that's not forgetting that when food is wasted, so is the energy, land, and resources that were used to create it.

Whether it's through poor storage at processing plants or uneaten food from our own kitchens, it all trickles down into an inefficient system that not only results in poor distribution, but also happens to be wreaking havoc on our planet.

After all, wasted food isn't just a social or humanitarian concern—it's an environmental one. Just like cows produce methane (a greenhouse gas emission even more potent than CO_2), so does our rotting food when it hits the landfill. According to experts, tackling food waste as a planet could eliminate up to 8% of our total emissions.

No doubt taking action against food waste could be a key step toward a more sustainable future and even provide better food security for billions around the world. Most importantly, though, the fight can start right at home. Let's take a look at some progress we've been making!

COMMUNITY FRIDGES

Community fridges are helping feed millions of people across the UK!

A community fridge is a space where surplus food is shared for free. This food comes from local businesses sharing food that won't be sold in time or from local gardeners and households.

Hubbub has set up the world's largest community fridge network with over 300 community fridges across the United Kingdom. This network saved over 7.5 million meals from going to waste in 2021.

As well as improving access to food, these fridges help empower communities to connect with each other and learn new skills through activities such as cooking lessons and workshops on how to grow your own food.

So if you need a hand with getting free food, have spare food to give, or even want to set up a community fridge in your neighborhood, check out how to get involved at the end of this chapter. There are also options for those not based in the United Kingdom too!

TACKLING FOOD WASTE

UK organizations are getting better at fighting food waste with food redistribution reaching a record high in 2021!

Food redistribution is key to minimizing food waste at the industry level. This will help not only to minimize the carbon impact of the food production industry, but will also help fight the cost of living crisis by helping to feed lower income households.

Since 2019, each year has seen an increase in the amount of food that has been redistributed in terms of weight. According to the annual figures on surplus food redistribution in the United Kingdom published by WRAP in 2021, 106,000 tons of surplus food was redirected, which amounts to an estimated value of $410 million (£330 million). This represents a 16% increase from the previous year.

The pandemic fostered a broader sense of social responsibility, which saw industries that were previously notorious for wastage shift toward redistribution schemes that have continued postpandemic.

Despite these improvements, WRAP, a climate action non-governmental organization (NGO), estimated that there were still 200,000 tons of consumable food that went to waste in 2021.

Experts warn that these positive patterns of redistribution are not guaranteed to continue, so it's important that we keep talking about it and keep requiring these industries to do their part.

Let's work together to make sure that it's only onward and upward from here!

SCRAPPING BEST-BEFORE DATES

Supermarkets in the United Kingdom are scrapping best-before dates to fight food waste!

Food spoilage is the number one source of food wastage from homes across the world. Two-thirds of purchased food is thrown away because it wasn't used by its best-before date. This amounts to 4.5 million tons of food per year! This means that all the energy and resources which went into making this food were in vain.

Supermarkets in the United Kingdom, like Waitrose, are aiming to reduce this waste by scrapping best-before dates altogether. They believe this will encourage the public to be more connected to the actual state of their food and allow them to decide for themselves whether food is still good enough to eat.

It is estimated that this plan could save up to seven million shopping baskets worth of food each year. This will not only help reduce the carbon impact of the agricultural industry, but also lower costs for consumers by reducing the frequency they have to purchase these items.

While some places are scrapping best-before dates, which are a guide for food quality, they are still employing sell-by dates on some products, like dairy items, which has more to do with food safety. Some supermarkets are encouraging people to use a sniff test for milk to use their own judgment before dumping it.

Hopefully we will see this plan implemented more in the future, so it can continue to play its part in reducing the climate crisis.

REROUTING FOOD FROM LANDFILLS

Food Forward, a California-based nonprofit, is on a mission to reduce food waste while simultaneously supporting those struggling with food insecurity!

So far, it has managed to stop a whopping 290 million pounds (132 million kilograms) of food from entering landfills. Instead, it donates the food to households in need, providing enough produce to feed 150,000 individuals their five daily servings of fruits and vegetables.

Founded in 2009, Food Forward works to recover food from farmers markets, produce markets, and even public orchards and backyard fruit trees, to help reduce the billions of pounds of food that are wasted in the United States every year.

It makes sense; food waste is terrible for the planet. When it rots in landfills, it produces methane, a potent greenhouse gas. But it doesn't need to be wasted, because according to the nonprofit, one in nine people in California do not have adequate access to food.

Food Forward recovers produce in 12 Southern California counties, along with six adjacent states and tribal lands to help reduce millions of pounds of food that is wasted every year.

STEPS YOU CAN TAKE

visit sambentley.co.uk/gnpe/food to access relevant links

O Get involved with a local community fridge! If you're in the United Kingdom, simply visit Hubbub and scroll around the map to find your closest community fridge. You can volunteer at one in your free time, or you can even set up your own fridge in your community. If you're not in the UK, you can check out websites like changeX, which also has a community fridge network, and Freedge, too. Both websites have maps and contact details on how to get involved!

O Plan ahead for meal times. Shopping with a list of what you need for the week can make a huge difference when it comes to making sure you haven't overspent on fresh ingredients that won't last. You'll even save money in the long run!

O Get creative! If you've got leftovers that you're not sure what to do with, don't throw them away; look around online or think outside the box for some inspiration to make the most of the ingredients you have in your cupboard.

O Follow @stopfoodwaste.ie on Instagram for tips and tricks about how to make the most of the food in your cupboards or when to buy the best seasonal fruits and vegetables that'll be fresher and kinder for your budget.

O Donate products you won't use to your local food projects that help the community and reduce waste at the same time. If you're based in London, check out FoodCycle, they've salvaged 191 tons of food waste in 2021 alone.

CHAPTER 9

Dairy & Meat

Leonardo DiCaprio is known for many epic movies, but there's one project he got involved in that seemed to fly under the radar. *Before the Flood* is a 2016 documentary made in conjunction with *National Geographic* that encourages viewers to make the transition to a more sustainable society, and there was one section from it that changed my life forever.

Leo is asking Gidon Eshel, PhD, a research professor, about the devastating impact of meat and dairy on our planet and asks, "So even cutting the amount of beef in half, or by a quarter, could make a significant difference?" Gidon responds, "Very significant. Even if you just switch to chicken, you will have eliminated 80% of what you emit, depending on where you're coming from."

From that point onward, I tested myself to see if I could cut out beef entirely and shift to just chicken and fish. It was surprisingly easy, so I carried on and tried being vegetarian for a month. A month became 2 years, and over that time period, I probably consumed more cheese than I ever had before. But as I learned more and more about the ways animals are treated in these systems, through documentaries like *Cowspiracy* and *Dominion*, I couldn't turn a blind eye to the abuse that still occurs in the dairy industry to these sentient animals that have the capacity to fear and love. So I became vegan.

Avoiding meat and dairy products is the single biggest way to reduce your environmental impact on the planet, according to researchers from the University of Oxford. The author of one study, Joseph Poore, told *The Guardian*, "A vegan diet is probably the single biggest way to reduce your impact on planet Earth, not just greenhouse gasses, but global acidification, eutrophication, land use, and water use. It is far bigger than cutting down on your flights or buying an electric car."

No other consumer products need as much land to produce them as meat and milk do. 77% of Earth's agricultural land is allotted for raising animals or tending crops to feed those animals. In Brazil, 432 million acres (175 million hectares)—equivalent to the agricultural area of the entire European Union—is used to raise cattle. More than 80% of farmland is used for livestock, but it produces just 18% of food calories and 37% of protein.

Cattle raising is the main driver of the destruction of ecosystems and the habitat they provide to indigenous peoples and traditional communities. In fact, if cattle were their own nation, they'd be the world's third-largest emitter of greenhouse gasses.

Animals are often kept in cramped and filthy conditions, and they may be subjected to painful procedures like tail-docking and debeaking. There is also growing evidence that the meat-and-dairy industry is a major contributor to antibiotic resistance, a prominent threat to medicine globally.

An Oxford Martin School study found that food-related emissions could be reduced by 70% if the world's entire population adopted a vegan diet (63% by adopting a vegetarian diet, and 29% by adhering to global dietary guidelines).

There are health benefits associated with transitioning to a plant-rich diet, too. The same study found that a global switch to diets that rely less on meat and more on fruits and vegetables could save 5.1 million deaths per year by 2050; with even greater benefits coming from vegetarian diets, avoiding 7.3 million deaths; and vegan diets, avoiding 8.1 million deaths.

Honestly, if I write another book, maybe it'll be about this topic, as I could go on, but for now, here's some good news stories in this space and some next steps you can take to find out more!

FARMERS GOING VEGAN

Farmers For Stock-Free Farming (FFSFF) is a grassroots initiative that's helping Scottish farmers move away from animal agriculture!

The organization argues that it's more difficult than ever for meat and dairy farmers to turn a profit (due to Brexit, for example), while also stressing how the industry is harmful to the environment. Cows, sheep, and goats all emit methane, which is 25 times as potent as carbon dioxide at warming the planet.

So FFSFF helps people like Laurence Candy, a former dairy and beef farmer, to grow crops, like oats and wheat. After all, the demand for plant-based alternatives is growing. In 2021, a report by Mintel noted that one in three Brits regularly drink oat milk.

The initiative also advocates for reforestation and encourages farmers to make money by renting their land out for camping or glamping.

Candy isn't alone. A number of farmers around the world are moving away from animal agriculture. In 2019, the documentary *73 Cows*, which followed an English beef farmer as he gave his cows to a sanctuary, won a BAFTA.

MEET THE CATTLE RANCHER WHO BROKE TRADITION

Richard Traylor, a sixth-generation Texan cattle rancher, and his wife, Cindy, went vegan in 2018, decided to stop farming animals, and moved toward sustainable, plant-based farming!

The pair made the decision after one of their cows, Honey, became injured. They were looking for a sanctuary for Honey, and came across the Rowdy Girl Ranch. The ranch was founded by Renee King-Sonnen, who used to be married to a cattle rancher before she went vegan and started her own sanctuary.

King-Sonnen had a big impact on the Traylors. After they met her, they also decided to give up eating and farming animals.

In 2020, the pair spoke at a summit for the Rancher Advocacy Program, which was also founded by King-Sonnen and aims to help farmers transition away from animal agriculture. Traylor said at the summit, "I used to laugh at veganism, and now it's a way of life."

A TOWN IN EUROPE MOVES TOWARD VEGANISM

In 2022 Haywards Heath, a town in West Sussex, UK, became the very first town in Europe to sign the Plant Based Treaty!

Intended to accompany the Paris Agreement (a legally binding international treaty focused on tackling the climate crisis), the Plant Based Treaty is a grassroots initiative.

Essentially, it's striving to create a plant-based food system, as the current food system relies heavily on environmentally damaging animal agriculture to produce foods like meat and cheese. The treaty's demands include no building of new animal farms and no conversion of any land for animal feed production.

Residents in Haywards Heath do not have to follow the treaty's principles, but the town council will now encourage members of the local community to consume more plant-based foods, which research suggests are far better for the planet and limit food waste.

Other towns and cities that have endorsed the Plant Based Treaty include Buenos Aires, Argentina; Rosario, Argentina; Mundra, India; and Boynton Beach, US.

THE PROS OF PLANT-BASED MEATS

Researchers have determined that plant-based meats are far better for the planet!

In August 2022, researchers from the University of Oxford examined 57,000 products from supermarkets in a bid to figure out which food items are best for the planet. They examined several factors, including land use, water use, and greenhouse gas emissions, to come up with their findings.

The researchers discovered that while dried beef items (like jerky, for example) are bad for the environment, plant-based meat products are among the best. In fact, the study suggests that vegetarian sausages and burgers are up to 10 times better for the planet than their meaty counterparts. The findings also note that foods made with fruit, sugar, flour, and vegetables have a low impact on the planet.

Professor Peter Scarborough, an Oxford professor who worked on the research, said that the study could be used to encourage retailers to reduce the environmental impact of their products, as well as to create some sort of tool that would help supermarket shoppers make better choices for the planet.

OTHER MEAT ALTERNATIVES

Switching red meat for mushrooms could help slow deforestation!

Deforestation is terrible for the planet. Not only does it disrupt wildlife habitats and vital ecosystems, but it also releases greenhouse gasses. In 2021, scientists revealed that the Amazon rainforest, which has been ravaged by deforestation, now emits more carbon dioxide into the atmosphere than it absorbs.

One of the biggest drivers of deforestation in the world is the beef industry. But if we replace beef with microbial protein products, which are typically made from fungi, algae, or bacteria, we could reduce deforestation significantly, according to new research.

The study, conducted by the Potsdam Institute for Climate Impact Research, notes that because microbial protein is produced in bioreactors (large vats) it requires far less land than traditional beef production. In fact, researchers concluded that if we replace just 20% of global beef consumption with microbial protein over the next 30 years, deforestation could be halved.

BANNING MEAT ADS IN PUBLIC PLACES

From 2024 onward, Haarlem, a city in the Netherlands, will no longer allow companies to advertise meat products in public places!

This is because the meat industry is harmful to the planet. According to the United Nations, it accounts for 14.5% of all global greenhouse gas emissions. It's also a leading driver of deforestation and habitat destruction, among many other environmental issues.

Haarlem's new rules will mean that, in 2 years, buses, shelters, and screens on the street will no longer show any advertisements for meat products. The city is the first in the world to enact such a ban. It was drafted by GroenLinks, a green political party in the Netherlands. In addition to meat, the city will no longer allow advertisements for flights or cars that run on fossil fuels, in a bid to encourage citizens to make better choices for the planet.

STEPS YOU CAN TAKE

visit sambentley.co.uk/gnpe/food to access relevant links

○ As Gidon Eshel, PhD says in *Before the Flood*, "If you want something you can do, without appealing to any higher authority, such as the government, I can't think of an easier out than changing your diet. You can start tonight!" Choosing a diet that is less impactful on the planet is something we can do multiple times each day. Shifting to a plant-based diet is one of the best ways to reduce your environmental impact on the planet.

○ Get yourself comfy on the sofa and learn more about the issues. A documentary is a great place to start. In the past few years, a number of films (like 2014's *Cowspiracy*, 2017's *What the Health*, and 2021's *Eating Our Way to Extinction*) have covered these topics in depth. After reading this chapter, also be sure to watch *Before the Flood*, which I found particularly inspiring.

○ If you need inspiration for some amazing meals that don't include meat and dairy, check out these amazing vegan chefs on YouTube: Cheap Lazy Vegan, Avant-Garde Vegan, and Rachel Ama are a few favorites, but there are so many! Me and my mum love the BOSH! books, too!

○ Read up on plant-based nutrition. A plant-based diet—rich in fruits, vegetables, grains, and nuts—is one of the healthiest ways to eat. But before you dramatically change your diet in any way, you need to do some research on nutrition. (Many vegans choose to supplement B_{12}, for example.) The Vegan Society, NHS (National Health Service), and Healthline are good go-to resources to check out.

CHAPTER 10

Composting

First off, what is composting, and why is it good for the planet?

Composting is the breakdown of organic material that creates matter (compost) rich in nutrients and microorganisms.

One-third of food produced globally goes to waste. When food waste goes to landfills, it's broken down by bacteria without oxygen present, which produces methane that is harmful to the planet.

Composting is an aerobic process and methane-producing microbes are not active in the presence of oxygen. So composting reduces your methane footprint; how cool is that!

Here's a little guide about composting if you want to get started at home.

The table below shows different methods of composting:

	INSIDE	OUTSIDE	MAINTENANCE	COMPLETION TIME
COLD		✳	LOW	6-12 MONTHS
HOT		✳	MEDIUM	4 WEEKS
VERMICOMPOSTING	✳	✳	MEDIUM	2-3 MONTHS
BOKASHI	✳	✳	LOW	2 WEEKS
COUNTERTOP DEHYDRATOR	✳		LOW	4 HOURS

How do the costs compare for various composting methods?

TYPE	EQUIPMENT COST	ONGOING COST
COLD	$$ (BIN)	
HOT	$$ (BIN)	
VERMICOMPOSTING	$$ (BIN & WORMS)	
BOKASHI	$$ (BIN & BRAN)	$ (BOKASHI BRAN)
COUNTERTOP DEHYDRATOR	$$$ (BIN, PODS, FILTER)	$$ (PODS & FILTER)

What materials can you add with each method?

	GREENS	BROWNS	COOKED/BAKED	PROCESSED	MEAT/FISH/DAIRY
COLD	✳	✳			
HOT	✳	✳			
VERMICOMPOSTING	✳	✳			
BOKASHI	✳	✳	✳	✳	✳
COUNTERTOP DEHYDRATOR	✳	✳	✳	✳	✳

Let's talk about pH Balance!

Finished compost has a neutral pH (between 6 and 8), and achieving this comes down to the materials you add and how often you turn the compost.

The general rule of thumb is to add 2 parts brown to 1 part green. Don't worry too much about achieving the perfect pH, focus more on adding a variety of materials and maintaining the correct level of moisture (similar to a damp sponge).

Composting food and yard waste helps reduce the amount of waste you produce and also helps create "black gold" soil in your garden, allowing more delicious foods to grow!

Let's see what other good news about composting has been happening around the world!

What materials can you add to your compost?

GREENS (NITROGEN-RICH)	BROWNS (CARBON-RICH)	BOTH	DO NOT ADD (TO HOT/COLD/VERMI)
FRUIT & VEG SCRAPS	PAPER / CARDBOARD	EGG SHELLS	COOKED FOODS
GRASS CLIPPINGS	DEAD LEAVES / PLANTS	PET BEDDING	PROCESSED FOODS
PLANTS / NETTLES	HAY / STRAW	OLD COMPOST	BAKED GOODS
UNSEEDED WEEDS	WOOD CHIPPINGS		MEATS / FISH
COFFEE GROUNDS	UNTREATED SAWDUST		DAIRY PRODUCTS
MANURE (HERBIVORES)	TWIGS / SMALL BRANCHES		MANURE (CARNIVORES)
PLASTIC-FREE TEABAGS	NAPKINS		
SEAWEED	PAPER COFFEE FILTERS		
URINE	WOOD ASH		
HAIR			
BREWERY GRAIN			

REQUIRING FOOD WASTE TO BE COMPOSTED

California passed a bill in 2016 to change the way residents and businesses dispose of waste, and its been in effect since 2021!

The new law dictates that leftover food scraps must be separated from other waste. In 2022, many households across California are receiving notices with instructions on how to correctly separate trash.

If they don't comply, people could face fines of up to $100 for a first offense, and up to $500 for repeat offenses. (Although these fines won't be coming into effect until 2024.)

The new law isn't designed to add to a Californian family's list of chores. It has been passed for environmental reasons, to prevent organic waste from ending up in the landfill. This is because when food is thrown out, it rots and emits methane. Luckily, composting food scraps significantly reduces these emissions.

TURNING BODIES INTO COMPOST

In 2020, Recompose, the first funeral home in the United States to compost bodies, opened for business!

Founded by Katrina Spade, an architect and death-care advocate, the funeral home essentially turns people into compost. It does this by laying bodies inside a vessel with wood chips, straw, and alfalfa for a period of 30 days.

The process is more environmentally friendly than traditional after-death rituals; cremation emits greenhouse gasses, while ground burial is resource-intensive and pollutes the soil.

In fact, Recompose states that for every person who chooses human composting, which boosts soil health, one ton of CO_2 is saved. Bodies composted with Recompose can also be donated to Bells Mountain, a legally protected forest in Washington state, which has suffered for decades from degraded soil.

Despite its eco-credentials, composting bodies isn't legal everywhere. However in 2021, both Colorado and Oregon joined Washington and legalized the practice. And in early 2022, it was revealed that California will allow human composting starting in 2027.

BANNING PEAT-BASED COMPOST

The United Kingdom is banning sales of peat compost to help protect natural peatlands that are under threat!

Around the world, many household gardeners use peat-based compost on their plants. This is because peat, which is made up of partly decayed plant matter, is great at retaining water and nutrients.

As of 2024, sales of peat compost will be largely banned in England, and it's for a really good reason: peatlands (wetlands where peat builds up over thousands of years) need to be left alone.

The UK is home to almost 7.4 million acres (3 million hectares) of peatlands, which cover about 10% of the country's land area. They have many environmental benefits, but they are particularly vital for wildlife habitats and carbon capture. In fact, in the UK alone, they harbor more than 3 billion tons of carbon.

But peatlands are under threat, not only from the gardening industry, but also from agriculture and hunting grouse, as gamekeepers burn peatlands to create different heather habitats, which are nutritious and can provide a place for nesting grouse.

While many conservationists have praised the peat-compost ban, which only impacts those buying it for private gardens or allotments, they also maintain there is much more to be done to protect peatlands.

NEW YORK'S COMPOST CHAMPION

One person is making a big difference to get more folks to compost!

Every year, New York City produces around 4 million tons of food waste. Most of this goes to the landfill, where it emits huge amounts of methane into the atmosphere. But there is a way to sustainably get rid of waste: composting.

Domingo Morales is the founder of Compost Power, a New York–based organization that aims to provide public housing residents with education and access to sustainable composting.

Morales grew up in public housing, and first learned all about the importance of composting food scraps with Green City Force, a nonprofit that trains young people for sustainable careers. Morales quickly became a standout member of the team, and in 2020, he was awarded $200,000 from The David Prize, an initiative that awards extraordinary New Yorkers with money toward their plans to improve the city.

Morales used the money to create Compost Power. Now, he builds new community composting sites, runs educational workshops for children, and provides training for young people.

STEPS YOU CAN TAKE

visit sambentley.co.uk/gnpe/food to access relevant links

O Based on where you live (with or without a garden), narrow down your options on how best to compost and then choose a method that suits your diet and what you typically throw away. There is an abundance of resources online to learn from such as Help Me Compost that has accessible guides, such as one called "Brown to Green Compost Ratio (The Easiest Guide Ever!)"

O If you live in an apartment and have no use for compost, use the free app, ShareWaste, where you can donate materials or compost. You can also see if a neighbor or local community garden can use your compost, too.

O To learn how to make your own compost bin and get more compost tips, follow Positively Green on TikTok and Instagram (@positivelygreenliving) or Huw Richards (@HuwRichards) on YouTube. I've used both and they've been super informative!

O Support Domingo and the amazing work he does with Compost Power by following his journey on Instagram (@compostpower) and checking out Compost Power NYC. It builds out community compost sites across New York City with an emphasis on underserved and marginalized communities. If you're in NYC, even better!

CHAPTER 11

Water

For many of us, access to freshwater is almost a given; we bathe in it, drink it, wash our dishes in it, and watch it flow down our drains with little thought about where it came from. But did you know that freshwater makes up just 3% of Earth's water, with just 1.2% of that water actually available for human consumption?

The rest is sealed up in the form of glaciers, permafrost, and ice caps, or lodged under the surface of the earth, which means more than 99% of the world's water is completely unusable for most living things.

Today, the water we drink has likely been around since the age of dinosaurs. While the amount available has remained steady, the population hasn't. Now, we're facing a global water crisis, and it's estimated that by 2025, two-thirds of the global population will reside in regions where water will become more scarce as a result of increased use, population growth, and climate stress.

In fact, extreme weather events, such as flooding, wildfires, and droughts, are a big part of the problem, threatening people's access to clean, sustainable water sources and often hitting the developing world the hardest.

The good news is that by putting water conservation at the heart of our action plans and investing in better infrastructure, we can build more resilient communities and fight climate change at the same time.

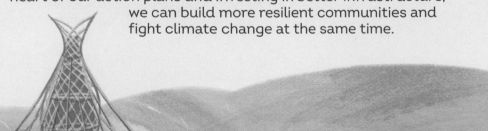

TURNING SEAWATER TO DRINKING WATER

Researchers at MIT created a portable device that can help make drinking water from seawater!

Right now, 785 million people around the world do not have basic access to safe drinking water, according to the World Health Organization.

But there is a huge body of water we have access to: the ocean. The only thing is, seawater is not safe to drink because it contains too much salt. It's not easy to remove the salt, as it requires huge amounts of energy to filter. However, scientists are working on solutions, and they're making progress.

Researchers from the Massachusetts Institute of Technology (MIT) have developed a desalination (salt removal) unit that relies on electrical power instead of filters and weighs less than 22lb (10kg). Plus, it requires no more power than a phone charger to operate and could be powered by solar.

These units alone can't solve water scarcity right away, but the technology could indicate what is to come in terms of providing more safe drinking water for all.

CLOUDFISHERS

CloudFishers are helping provide super dry regions with clean water!

In many parts of the world, people don't have access to clean water and rainfall is very rare, but if there are lots of clouds and fog, these are like natural wells in the sky!

So let me introduce you to the CloudFisher, which can be used to take water out of the sky and give it to the people. How does the CloudFisher work? When fog starts to form in an area, wind drives the fog into its nets. Tiny droplets of water get trapped in the mesh and merge into larger droplets, heavy enough to move down the nets into a collecting trough.

This water then travels down beneath the nets to storage tanks, which makes the water ready for use. This water can help produce food and empower local people to create their own urban farms or be used for reforestation projects. According to a BBC feature on cloud nets in Peru, the nets gather around 50 to 100 gallons (200 to 400 liters) of water a day!

Since 2007, the German Water Foundation has been using fog collectors to obtain drinking water. Their CloudFisher innovation marked a breakthrough in sustainable fog harvesting technology and is the world's first production fog harvester capable of withstanding wind speeds of up to 75 mph (120kph).

This is important as CloudFishers require specific conditions to be effective, such as consistent fog and light winds. There also needs to be land available to install them on. While they may not be viable in many locations, in the communities they can and do serve, they provide a vital source of water supply. They are particularly effective in coastal areas, where there are high amounts of fog and large quantities of fresh water in the air.

In Sub-Saharan Africa alone, women or girls spend around 40 billion hours a year fetching water for their families, which disrupts their ability to go to school, maintain jobs, or go about their lives unencumbered by the burden of water scarcity. Solutions like this one can help address existing inequalities and support vulnerable women, too!

WARKA WATER TOWER

A magnificent tower can capture clean water from thin air and help provide remote communities with 26 gallons (100 liters) of water every day!

It's called the *Warka Water Tower*, inspired by the warka tree that's native to Ethiopia. The tower collects moisture from the air and brings it back down to the people. It can collect rainwater, harvest fog, and capture dew.

The structure is made from bamboo with an inner mesh lining that harvests and collects droplets of water by condensation. These droplets then trickle down the mesh, fall through a big funnel, and drop into a reservoir that holds onto the water. The base of the tower is kept in the shade by a canopy that helps prevent the water from evaporating. So far, the Warka Water Tower has been implemented in Ethiopia, Haiti, Brazil and many more places!

An amazing part of the project is that it's an open source concept, meaning its design is publicly accessible, so local communities can be trained on how to create it and make as many as they like. In Ethiopia, their estimated cost per tower is only around $1,000!

Around 1 in 10 people don't have access to safe water, so solutions like this are really important to help transform millions of lives for good.

PLUCKING DRINKING WATER FROM DESERT AIR

A low-cost gel film made from abundant materials has been invented that can capture potable water from desert air!

More than two billion people around the world live in drylands, which are areas where water is scarce. For example, 60% of the African continent is dryland, and this means that the populations who live there are vulnerable to water shortages.

Thanks to recent innovation, there could be some hope right around the corner. Using cellulose and konjac gum, University of Texas researchers have developed a gel that literally pulls water out of the air, even in places where the climate is dry.

The solution is cheap (at only $0.9 per lb/$2 per kg) and effective. Just one kilogram of the gel can result in more than 1.6 gallons (6 liters) of water per day. Further research could increase this even more.

In a bid to assist soldiers working in dry climates, the research was funded by the US Department of Defense. But if it continues, the innovation could help many other people around the world, too!

STEPS YOU CAN TAKE

visit sambentley.co.uk/gnpe/food to access relevant links

○ Think about water usage around your home; only wash full loads of laundry, turn off the tap while brushing your teeth, and make sure to fix any leaks. Even taking a shower instead of a bath can save more than 11 gallons (40 liters) of water in one go!

○ Save water in the garden with a rain barrel. Sprinklers can use up to 264 gallons (1,000 liters) of water an hour, while a rain barrel can catch large amounts of rainwater that can be used for watering plants, and washing cars and windows. The switch is a no-brainer.

○ Check out @saveourwater on Instagram, a state-wide program aimed at helping Californians reduce everyday water use through conservation ideas, inspiration, and tips. These tips can easily be transferred into the everyday lives of people living around the world.

WILDLIFE

Did you know that species are going extinct 10,000 times faster than the natural, historical extinction rate?

In a world that has seen the human population boom over the last 200 years, natural resources are being consumed at an alarming rate, without a chance for recovery—and our global wildlife is feeling the brunt of it.

In fact, in less than 50 years, wildlife populations have declined by more than two-thirds, as forests are burning, oceans are being overfished, and wild areas are being destroyed by human activity.

Wildlife makes up a vital part of the world's ecosystems, providing balance and stability to nature's processes. From the bugs and bees that pollinate plants and play a vital role in food production, to the large mammals who, as it turns out, can help in the fight against climate change, it's never been more important to protect our wild species.

And it's not just limited to how wildlife can support our food and ecosystems either. Studies have shown how beneficial wildlife is to our mental health, too, with evidence showing that those who spend more time in nature are happier and less stressed than those individuals who don't get outside as much.

Thankfully, the world is listening and things are slowly changing. More funding is available than ever before to support vital projects working to limit the human impact on wildlife and their ecosystems, while an increasing number of us are educating ourselves and becoming inspired to take action for things like animal rights and conservation efforts in our own ways. Honestly? There's no better time to take action!

CHAPTER 12
Reintroducing Species

One of the main ways we can support wildlife and promote biodiversity at the same time is through reintroduction programs. These programs work to reintroduce a species into an area that once thrived there but has since disappeared.

There are a number of reasons why a species may no longer live within its native area, from overexploitation and habitat loss to an invasive species taking over. But often, there's a very important reason the species was there in the first place, acting as a part of the natural process that supported a wider ecosystem. For example, Britain's reintroduced beavers are helping reduce flooding and boosting wildlife populations, while the gray wolves in Yellowstone National Park have been improving the health of the landscape since their reintroduction in the 90s.

It's not an easy process, and most of the time, reintroduction starts with a lot of research, including coming to an understanding of why the species originally disappeared, what the impact could be when the species is brought back to the area, and how to choose the correct individuals for release. Once the species is reintroduced, it's important to be constantly monitoring the newly brought-back wildlife in the habitat to evaluate and understand its local impact.

Today, scientists and ecologists are seeing more success than ever before with reintroduction programs, as more money is pumped into conservation and environmental planning. And that's great news! Why? Because, by reestablishing these natural processes and restoring ecosystems, we can start to rebuild the world's life support system and tackle climate change and species extinction, all while restoring our land and seascapes.

EUROPEAN BISON REINTRODUCTION

If you visit the Dutch coast today, you may stumble across the European bison!

Hunted to extinction almost 100 years ago, a reintroduced herd of European bison is now thriving in the Kraansvlak coastal dune reserve.

In fact, since the program started back in 2007, the project has been so successful that European bison have now been introduced to other natural areas in the Netherlands, as well as other parts of Europe. They've even shaken off their vulnerable status and moved to near threatened on the IUCN Red List of Threatened Species.

But why are they being reintroduced? Well, European bison are actually categorized as a keystone species, which means that the greater health of their ecosystem depends on their survival. From managing vegetation to dispersing seeds and nutrients across their territory, studies are showing the positive impact of their reintroduction. The natural management of shrubs and grasses is just one example of this.

Wild bison have also been returned to the United Kingdom for the first time in thousands of years. In July 2022, the Wildwood Trust and Kent Wildlife Trust launched a flagship wilding project called *Wilder Blean* and introduced three European bison to help transform a dense pine forest into a thriving natural woodland.

An earlier species of bison used to roam the UK thousands of years ago, carrying out all these important functions in woodlands before it sadly went extinct. The European bison was selected as it is the closest living relative and provides the best chance at recreating those grazing behaviors that once existed.

Although the concept of a wilding project is to take a hands-off approach and give nature the freedom and space to do its thing, rangers are in place who will look after the bison, giving them regular health checks and ensuring they're safe.

RED & GREEN MACAWS

Macaws are getting a second chance in Argentina, a country where the bird has been extinct for more than a century!

Today, a big rewilding project is working to bring macaws back to the Corrientes Province of Northeast Argentina.

Launched by Rewilding Argentina in 2015, the initiative is all about restoring some of the native fauna in Iberá National Park, to help support the protected wetlands and save the dwindling Paraná Forest.

The red and green macaws are seen as a crucial part of the recovery process because as seed dispersers, they can help to regenerate native species in a natural, sustainable way.

So far, around 20 birds (15 in 2015 and another 5 in 2020) have been released into the area, and there's evidence of breeding already, with the first successful wild chicks born in the wetlands in 150 years.

They're not the only part of the project either. Other species are also being reintroduced—including the giant river otter—as part of a sustainable development model built around the recovery of ecosystems and nature-based tourism.

STEPS YOU CAN TAKE

visit sambentley.co.uk/gnpe/wildlife to access relevant links

O Advocates for nature are needed now more than ever. If you were inspired by the bison reintroduction project, you can become a member of the Wildwood Trust or Kent Wildlife Trust or find a trust local to you and see how you can get involved.

O Supporting reintroduction programs can start right in your backyard! Research what projects are happening locally in your area to support native species, and find out if there are specific plants you can grow, or even small habitats you can build, to help the projects succeed.

O Check out the Beaver Trust's website to find out how they're reintroducing beavers to the United Kingdom to help build climate-resilient landscapes and restore freshwater habitats. Make a donation, fundraise, or spread the word to help this species become a key part of the nation's nature network once more.

O Follow the Sumatran Orangutan Conservation Program on Instagram (@socp.official) to discover how they're fighting to save one of the most endangered species in the world through reintroduction.

CHAPTER 13

Conservation Success Stories

Conservation is all about protecting our world for the future, whether that's through responsible use of natural resources, maintaining a diversity of species, or supporting ecosystems so they thrive.

But why is conservation so important? Over the last century, an unsustainable reliance on the natural world from the human population has taken a big toll on the health of our planet. This is now being seen through the ever-growing list of endangered species and the increasing impact of climate change—all of which links back to overconsumption and poor conservation management in the past.

Right now, conservation is one of the key ways we can take direct action to mitigate the negative impact, and even reverse some of the damage already done. And yes, there are some big threats that can hamper conservation efforts, which include everything from overfishing, deforestation, and the illegal wildlife trade, all resulting from human activity—but there is good news too.

More and more conservation projects are popping up across the world, and there is hope. From launching species reintroduction projects to countries coming together to save species from extinction, communities across the world that are working together toward creating a healthier planet. The best part? You can do it too!

WILDCATS

Wildcats are coming back to the United Kingdom!

Some great news came out of a conservation center in Scotland in 2021. Eight wild cat kittens were born at the Highland Wildlife Park, thanks to the team at Saving Wildcats who are working with experts to restore Scotland's critically endangered wildcat population.

On top of this, three rare snow leopard cubs were also born-named Padme, Maya, and Yashin, which was a nod to the species' native Himalayan home!

Part of Saving Wildcats' efforts include breeding and releasing these cats into carefully selected locations in Cairngorms National Park, which is home to over 25% of the UK's rare and endangered species.

These kittens are the future of wildcats in Scotland, and without the incredible work organizations like Saving Wildcats do, they would likely go extinct in Britain.

TIGER POPULATIONS

After more than a century of steady decline, tiger populations are finally on the rise!

A report from WWF showcased this rare and hard-fought conservation success story, which is a clear turning point in the history of tiger conservation.

In 2010, tiger populations hit an all-time low of around 3,200. Just a century ago, populations were estimated to be upwards of 100,000. To address this sharp decline, for the first time in history, all 13 tiger range governments met at the St. Petersburg Summit to collectively commit to doubling the wild tiger population by 2022—one of the greatest demonstrations of multinational collaboration to protect a single species ever.

The summit report states, "In saving tigers, we also save so much more. As apex predators, wild tigers are ecosystem controllers, keeping other carnivores and herbivores in check, which helps to maintain healthy vegetation and habitats. In turn, this supports invaluable ecosystem services for billions of people living in tiger landscapes—from clean air and freshwater to fuel and medicinal plants. Globally, tiger landscapes play a significant role in containing greenhouse gas emissions in their forests, grasslands, and soil, and buffering against the impact of natural disasters."

HUMPBACKS ARE BACK

After only 300 were left at one point, humpback whale populations are now booming in Eastern Australia!

Before commercial whaling was banned in the 1960s, around 99% of the Eastern Australian humpback whale population was decimated, according to Professor Mike Noad, director of marine studies at the University of Queensland. Sixty years after that ban, it's predicted there are around 40,000 humpback whales in the region.

The commercial whaling industry hunts whales for their meat and blubber, which could be consumed or melted down into oil.

"All we had to do was stop killing them; we haven't done much else apart from leaving them alone," Noad said, as reported by *The Guardian*. Now, local people and visitors can enjoy a longer whale watching season and watch these gentle giants thrive!

RIVER OTTERS

River otters have been spotted in the Detroit River for the first time in a hundred years!

The introduction of the fur trade in Detroit was unfortunately bad news for local beaver and otter populations, whose numbers were decimated as a result. As the city expanded, that growth came at the cost of these animals' habitats and safety.

However, the recent news that river otters are coming back has local people, and biologists in particular, feeling hopeful. The presence of otters and beavers usually signals the relative health of riparian habitats, so let's hope we see more of these animals returning to Detroit soon!

SUMATRAN RHINOS

A critically endangered Sumatran rhino was born at an Indonesian sanctuary!

Sumatran rhinos are the smallest species of rhinoceros, and most closely related to the extinct woolly rhinos. It's estimated there are less than 80 left on the planet, largely due to poaching and habitat destruction.

The news of this birth comes from the Sumatran Rhino Sanctuary, located in Way Kambas National Park in Indonesia. It's sparked massive hope for the future of the species with the total number of Sumatran rhinos in the sanctuary now growing to eight.

It's heartwarming to know we share a planet with these beautiful creatures, and that there are people working tirelessly to protect them from extinction.

GOLDEN EAGLES

The highest number of golden eagles has been recorded in the south of Scotland since the early nineteenth century!

Once on the brink of extinction, golden eagle numbers are soaring in Scotland, after the introduction of the South of Scotland Golden Eagle Project, which is helping reinforce the population growth of one of Scotland's most iconic species.

Only two to four pairs of golden eagles had been recorded across Dumfries and Galloway and the Scottish borders, but as of March 2022, the total number of golden eagles in the south of Scotland stands at around 33!

The timing of this project was critical to prevent the threatened, small population of eagles from vanishing from their native home, which was the case in England and Wales. The South of Scotland Golden Eagle Project aims to prevent the extinction of these eagles from Scotland's southern skies so they can once again thrive!

STEPS YOU CAN TAKE

visit sambentley.co.uk/gnpe/wildlife to access relevant links

O Check out Saving Wildcats' website to learn more about the incredible work they do. You can even sponsor a wildcat to help secure a prosperous future for this species in Scotland!

O Try creating a positive impact by volunteering with your local wildlife project or cleanup crew each week. Or why not use your vacation days to travel abroad and support global conservation initiatives at the same time? The Great Projects is a great place to get started.

O Use your voice for change. Organizations like Greenpeace work internationally and have all the tools you need to get out there and be heard by your leaders—from protesting against deep sea mining to petitioning your local government representatives.

O Stay educated by checking out conservation activist and writer George Monbiot on Instagram (@georgemonbiot) for regular content debunking environmental myths and highlighting real solutions.

O Donate, fundraise, and adopt animals from conservation charities that need help to keep their work going. You can be the help they've been waiting for!

CHAPTER 14

Animal Rights

The animal rights movement has never been louder or more engaged than it is today, but did you know that it actually goes back way further than many of us realize?

In fact, the view that animals are equal to humans has been around since ancient Egyptian times. Even great figures of the classical world, like the philosopher Pythagoras followed a vegetarian diet, and believed all animals should be treated as kindred spirits.

While the first laws around animal welfare started to come into play during the early 1800s, it wasn't until the 1970s that the animal rights movement really started to snowball with the publication of the book *Animal Liberation* by Peter Singer in 1975. Since then, there's been a growing movement by organizations and individuals on a global scale advocating for all animals to live free from exploitation and suffering.

From tackling the dairy, fishing, and meat industries, which not only cause massive suffering for animals on a horrific scale, but are also key drivers of climate change, to fighting fur in fashion, today's animal rights movement is making progress like never before.

SENTIENT-BEING CLASSIFICATIONS

In late 2021, the UK government announced that lobsters, octopuses, and crabs are sentient beings!

This was declared after a review by the London School of Economics and Political Science found there is strong evidence these species have the capacity for feelings. The review defined sentience as "the capacity to have feelings, such as feelings of pain, pleasure, hunger, thirst, warmth, joy, comfort, and excitement."

As part of the announcement, the UK government stated, "The scope of the Animal Welfare (Sentience) Bill has been extended to recognize lobsters, octopuses, and crabs and all other decapod crustaceans and cephalopod molluscs as sentient beings."

Vertebrates—animals with a backbone—are already recognized by the Animal Welfare Bill as sentient. However, even though invertebrates lack a backbone, some of them have complex nervous systems, one of the defining features of sentience.

The announcement will help take into account the welfare of these highly intelligent cephalopods and decapod crustaceans, and when the Animal Welfare Bill becomes law, experts in the field will form an Animal Sentience Committee that can issue reports on the impact of policy decisions on these species.

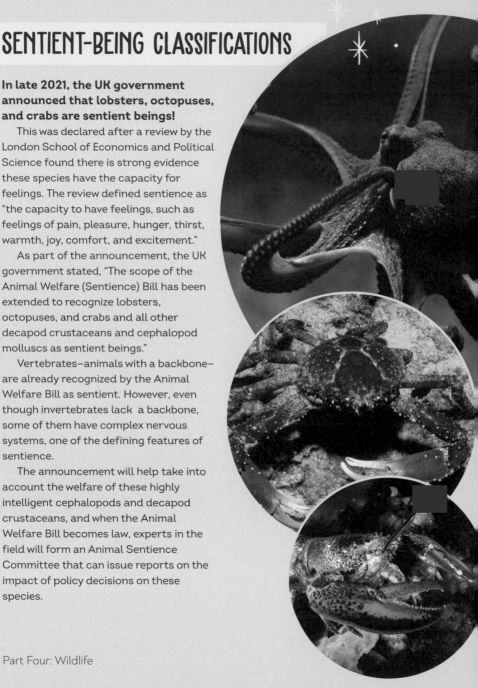

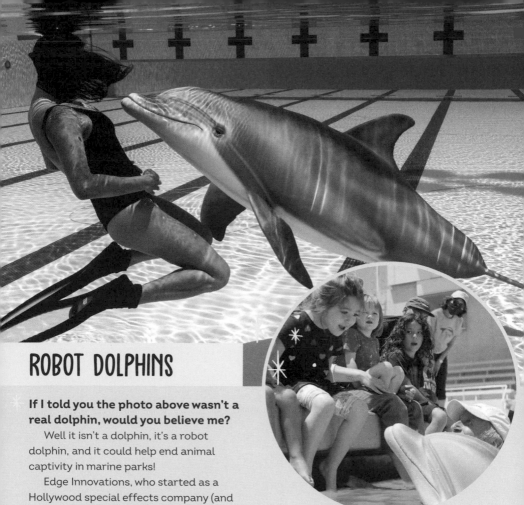

ROBOT DOLPHINS

If I told you the photo above wasn't a real dolphin, would you believe me?

Well it isn't a dolphin, it's a robot dolphin, and it could help end animal captivity in marine parks!

Edge Innovations, who started as a Hollywood special effects company (and even built Free Willy) developed the robot dolphin to show a cruelty-free way of interacting with marine life. When they tested the robot in an aquarium, even the fish seemingly believed it was real!

Though the dolphins are expensive to build, they are more economical in the long run than real dolphins and would save millions of dollars for marine parks over a 10-year period. They are currently human-operated, but the organization is developing a way for the dolphins to swim by using artificial intelligence alone.

These dolphins could help end cruel practices, like having photos taken with dolphins at marine parks or the inhumane performances dolphins are forced to participate in.

PROTECTING BEAVERS

Beavers will be legally protected in England from being captured, killed, injured, or disturbed!

The UK government announced this decision shortly after beavers were reintroduced to London after 400 years of absence. It will also be illegal to damage the habitats where they breed.

Tony Juniper, chairman of the government conservation agency Natural England, said, "This is a significant moment for beaver recovery, as we see a return of this species to its natural places in England. Beavers play important roles in making ecosystems function properly, helping to store and slow the flow of water through the building of dams and creation of complex wetlands, which in turn can reduce the impact of both floods and droughts, thereby assisting with adaptation to climate change. Those wetlands also catch carbon and aid the recovery of a wide range of other species. We are working closely with landowners, environmentalists, and other stakeholders to develop practical guidance to ensure these wonderful animals are able to thrive in suitable habitats alongside people across England."

It's great to see beavers being protected in the United Kingdom!

ECUADOR'S WILD ANIMALS

Ecuador has ruled that wild animals have the legal right to exist!

Located in the neotropics, bisected by the Andes mountain range, and buffeted by ocean currents that create unique ecosystems along its coastline, Ecuador is one of the 17 megadiverse countries in the world.

In 2022, a landmark decision was made by Ecuador's high court in a case involving a woolly monkey that ruled wild animals possess the legal right to exist, develop their innate instincts, and be free from disproportionate cruelty, fear, and distress.

The decision marked the country's first application of the "rights of nature" to a wild animal. "Rights of nature" is a legal theory which recognizes that our ecosystems, including trees, oceans, animals, and mountains, have rights just as human beings have rights.

It's reassuring to see courts around the world recognize these rights and start rethinking our relationship with nature, from one of dominance to one of stewardship and respect.

TEXAS BANS CHAINING DOGS

A law called the *Safe Outdoor Dogs Act* went into effect in Texas and banned chaining up dogs outside!

The law also states that dogs need to be provided with adequate shelter to protect them from harsh weather conditions and have access to drinkable water, too. Before the Safe Outdoor Dogs Act was passed, state law did not include these vital requirements.

One of the most significant changes in the act is that the 24-hour warning period for bad actors has been removed, meaning officers can take immediate action to help dogs in distress from now on, which is crucial for their safety!

BANNING HARMFUL ANIMAL PRACTICES

Italy has made progress in banning harmful practices that kill innocent animals!

Fur farming came to an end in Italy in 2022 after the Italian Senate voted to approve an amendment that forced the closing of all the country's mink farms.

Italy joined a list of countries including Japan, United Kingdom, and Germany that have already passed legislation banning this cruel industry from practicing within their borders.

While we're speaking about animal rights in Italy, it's important to note that Italy has also banned the killing of male chicks after two years of campaigning by Animal Equality Italy! This will end the cruel shredding of nearly 40 million male chicks a year, killed by the egg production industry because they are considered to be "nonproductive."

ENDING HORSE-DRAWN CARRIAGES

Mallorca has announced plans to ban horse-drawn carriages and replace them with electric alternatives!

This is thanks to the hard work of local animal rights activists at Progreso en Verde, who campaigned for years to scrap horse-drawn carriages in Palma, Mallorca, and pushed the council to approve a proposal to ban them by 2024.

There has been building pressure to ban horse-drawn carriage rides around the world, as more awareness is being brought to the suffering of horses that are forced to pull carriages filled with tourists through the busy streets and how it impacts their health.

A petition by PETA emphasized that horses can live up to 35 years when allowed to live a life free from exploitation.

Animal rights activists are continuing to push to end horse-drawn carriages around the world, with cities like Barcelona, New Delhi, Oxford, and Tel Aviv already taking that next step!

STEPS YOU CAN TAKE

visit sambentley.co.uk/gnpe/wildlife to access relevant links

O Take action to help ban fur farming around the world! Organizations like PETA have resources for activists who want to speak out, make a difference, and be a voice for the voiceless. Check out their website to learn more.

O Unite with groups like Animal Rebellion, that are working to fight the system from every level to change the world for animals. You can find out more about their campaign for a plant-based future and how to get involved on their website.

O Spread awareness online alongside the likes of Earthling Ed who you can find on YouTube and Instagram (@earthlinged) sharing facts, stories, and news about the reality of animal welfare and what we can do to change it.

O Lead by example in your everyday life. From what you wear to what you eat and the businesses you spend your money on, you can actively stand up for animal rights with every choice you make.

PART FIVE

CITY

Green living isn't always synonymous with those who live in cities. Let's be honest, pollution, poor land management, and overpopulation are often the first things that come to mind.

But did you know that bigger cities are actually better for the environment than their smaller counterparts?

Far from urban wastelands that overwhelmingly drain resources, bigger cities (that have been well-designed) offer the opportunity to reduce energy use and emissions, with better access to amenities like public transportation and education, while decreasing human encroachment on natural habitats.

In fact, while cities are often cited as the cause for issues like climate change, they can actually be part of the solution to the problem. Take, for example, an apartment building–it's going to be more efficient and cooler than a detached house and will take up less space than a single family home, which could encroach on natural habitats, agricultural land, and wildlife corridors.

As the world population grows, we're likely to see more people moving into urban spaces too. According to the UN, by 2050, 68% of the world's population will live in urban areas, so it's now more important than ever for us to take action and ensure these developed environments are as green as possible as they grow.

Most cities still have big issues to work through—including poor waste management and inefficient use of land—but from greenifying public spaces to developing urban farms, there are more green opportunities around city living than many may realize.

——————————————————

CHAPTER 15

Urban Farms

When you think of farming you'll likely picture sprawling fields of golden wheat, miles away from any city or urban landscape. Today, though, things are changing—and urban farming could be the way forward.

This concept is based on growing food in the center of urban spaces, whether in people's gardens, on their balconies, or via community-shared allotments. Basically, anywhere you can place a pot or till some soil you have the chance to join the urban farming revolution.

It doesn't stop there, though. The trend is now a growing industry with businesses beginning to invest in urban agriculture, using intensive indoor hydroponic or aquaculture facilities to grow masses of produce in much smaller spaces. The US government has even announced a new advisory committee dedicated to the practice due to the growing potential.

Why is this so important? Urban farming can be used to ease a number of issues in communities, including food security and even climate change. But at a more local level, with communities banding together to support each other through food growing projects, the mental health benefits can also be significant—with studies indicating that those who work in community gardens enjoy higher levels of resilience and optimism.

So is urban farming really the future? While it may not answer all the food problems in the world, it is without a doubt a step in the right direction.

GROWING FOOD IN VERTICAL FARMS

One of the world's largest vertical farms was launched in the United Kingdom!

The vertical farm is based in Lydney, Gloucestershire, and will have the equivalent of 70 tennis courts worth of growing space.

James Lloyd-Jones, founder and CEO of the Jones Food Company that is behind the development, said in a statement that they "plan to be able to supply 70% of the UK's fresh produce within the next 10 years." This would significantly cut down on the air and road miles food has to travel to reach the UK, while also making it less reliant on other countries, as around 46% of all food consumed in Britain is imported.

Food in a vertical farm can be grown entirely without pesticides and uses 95% less water. Risks such as fertilizer runoff—a major source of water pollution where excess liquid flows across the surface of the land, transporting fertilizer into nearby rivers and streams—can be avoided too.

While vertical farming is still a small part of the UK's agriculture industry, developments like this show the innovative ways the world is learning to grow produce!

NEW YORK'S MUSHROOM FARM

A huge vertical mushroom farm was opened in New York that has the capacity to produce nearly 3 million pounds (1.4 million kilograms) of mycelium annually!

In 2022, MyForest Foods launched the Swersey Silos, a 78,000-square-foot vertical mycelium farm, the world's largest farm of its kind.

If you're not familiar with mycelium, MyForest Foods describes it in this easy to understand way, "If you imagine a mushroom growing on the forest floor, you're likely picturing its cap and stem. Mycelium is the root-like structure found growing beneath the surface of the soil. It consists of oodles of thin filaments called *hyphae* that branch out in search of nutrients to consume."

The farm will be used to produce fungi-based bacon, an alternative to traditional pork bacon that drastically reduces the land, water, and carbon footprint found in conventional factory farming practices.

Eben Bayer, cofounder and CEO of MyForest Foods, grew up raising and slaughtering pigs on a family farm and developed MyForest Foods to help prevent the awful conditions and suffering pigs go through in the industrial pork industry.

By 2024, MyForest Foods is projected to serve their fungi-based bacon to more than one million consumers. That's a lot of water, emissions, and lives saved!

BRUSSELS' ROOFTOP GARDENS

Rooftop gardens are being brought to cities across the world!

In Brussels, Belgium they've begun trialing growing fruits and vegetables on the rooftop of a local supermarket.

The Lagum Project, financed by the European Union, aims to discover whether supermarket rooftop gardens can be a sustainable way to bring produce to local communities, while also helping to empower and support those in need.

In the first couple of months of the garden's launch, more than 2 tons of fruits and vegetables were harvested. If the trial is successful, these urban gardens will be grown on rooftops across Europe!

DETROIT BLACK FARMER LAND FUND

In Detroit, Black farmers are turning derelict properties into urban farms!

The Detroit Black Farmer Land Fund is helping make land ownership and healthy food more accessible, while also aiming to address the historical, racial land-ownership disparity that has affected Black farmers across the United States.

On Juneteenth 2020, three long-standing Detroit urban farming organizations came together to rebuild intergenerational land ownership for Black farmers in Detroit and to build a flourishing food-sovereign ecosystem within the city.

It's estimated that up to 40 square miles (104 square kilometers) of Detroit's land is vacant, out of the city's total 139 square miles (360 square kilometers). With the help of the Detroit Black Farmer Land Fund, Black farmers are turning these vacant lots into farms, providing an abundance of healthy produce to local people.

More funds are being launched to help support Black communities across the United States, using the Detroit Black Farmer Land Fund as a case study.

A HUGE URBAN GARDEN FOR BRAZIL

Rio de Janeiro will be home to the biggest urban garden in the world by 2024, coming in at the size of 15 soccer fields!

The Hortas Cariocas project, launched in 2006, promotes organic growing and provides a source of food and income to thousands of disadvantaged families in Rio. The project has been a massive success since its launch and has even helped transform some of the most deprived areas of Rio into areas of abundance and opportunity.

For example, green gardens transformed the vulnerable Manguinhos neighborhood in Rio—formerly occupied by drug users—into one of the largest community gardens in Latin America.

The project's objective is to keep expanding the gardens toward other areas with high rates of poverty and social exclusion, helping provide safer livelihoods for local people and more access to healthy produce.

Julio Cesar Barros, the founder of Hortas Cariocas, states that the goal is for each garden to become self-sustainable and ultimately independent. Gardeners donate 50% of the produce to people in need and the other 50% is priced at affordable rates and sold to the community.

STEPS YOU CAN TAKE

visit sambentley.co.uk/gnpe/city to access relevant links

O Read about how one mayor in Paris used the "15-minute city" concept and why the idea is spreading around the world. Based on improving quality of life, it's all about ensuring everything a person needs can be reached within 15 minutes by foot or bike.

O Start your own doorstep garden. Even if you don't have a lot of green space around your home, planted pots can make a real difference—your mental well-being and the bees will thank you for it.

O Follow Kyle Hagerty on Instagram (@urbanfarmstead) for tips and tricks on starting your own urban garden. You'll find tips on transforming abandoned lawns into thriving vegetable patches and more!

O Help address historical, racial land-ownership disparity and donate to the Detroit Black Farmer Land Fund. You can make a financial donation or donate gently used equipment and even land!

CHAPTER 16

Bee Habitats

During my childhood, I had a bad experience with a bee when I went to grab my bike's handlebar and interrupted a big bumblebee from having a little rest on it, and it gave me a nice painful sting in return. From that point on, I was always running away from bees, and this fear stuck with me for most of my life. It wasn't until I started looking into how amazing bees are, that my perspective started to change, and I fell in love with these little critters.

Bees pollinate a third of the food we eat and 80% of flowering plants. Strawberries, apples, almonds, and so many more wonderful foods all depend on the pollination of bees. Basically one out of every three bites of food we eat is thanks to bees!

Bees contribute massively to our economies, too. It'd cost farmers in the UK $2.2 billion (£1.8 billion) per year to manually pollinate their crops. In the US, the honey bee is responsible for $15 billion worth of agricultural crops each year. Bees and other pollinators are responsible for 35% of the world's crop production, as reported in 2019. However, agriculture is a driving force behind habitat destruction for bees. By the 1990s, an estimated 70% of deforested areas were converted to agricultural land. The common use of agrochemicals is also detrimental to bees.

Bees tend to suffer from innocent decisions we make at home. In the springtime, we mow the clovers and dandelions that early-emerging bee species forage. Many "bee-friendly" plants that we buy to help bees in our gardens are treated with harmful fungicides and insecticides. Healthy bee populations are crucial for our survival on this planet, so we must make it an international priority to protect and restore bee habitats before it's too late. Luckily, there are organizations that are doing just that. Let's take a look at some!

BEE BRICKS

A husband and wife created the Bee Brick, which can be used when building houses to create safe havens for solitary bees!

Cornish wildlife company Green&Blue designed a brick to help make homes into havens for bees. The Bee Brick is an innovative nesting site for solitary bees that can be used in place of a standard brick in construction.

They are the same size as regular bricks and replicate the natural habitats where solitary bees can nest. Looking after solitary bees is really important, as just under 250 of the 270 species of bees in Britain are solitary species, and they play a crucial role in the natural ecosystem.

In January 2022, it was announced that Bee Bricks would be compulsory on new buildings in Brighton, England that are taller than 16 feet (5 meters). With urban development as one of the key causes of habitat loss for bees, this move could help save thousands of bees, providing nesting and resting spaces for different species.

Though the bricks are made from concrete, they are composed of 75% waste material from the Cornish china-clay industry to help reduce the environmental impact. The business is also powered by 100% green energy and uses harvested rainwater in its production.

We need more planning regulations like this one that support and make space for nature. It's great to see Brighton leading the way on this as they do their part to help save the bees!

DETROIT HIVES

Derelict spaces in the Detroit area are being rejuvenated by the healing powers of local honey!

In 2016, a Detroit couple founded a nonprofit organization, motivated by the desire to transform these disused spaces, which if left to fester, would have fostered environmental hazards and given these areas a bad reputation.

The couple wondered if they could use local honey to help combat seasonal allergies that might also be exacerbated by overgrown ragweed in abandoned areas, and so Detroit Hives was born.

With the help of the Detroit Land Bank Authority, an agency that works to redevelop abandoned properties, they have been able to buy these unproductive lots and restore them to productive spaces. The team also travels around to local schools and hosts tours of the honey farm to build support and interest for their specific projects. Their goal is to instill a sense of care and responsibility for the ethical treatment of pollinators in urban settings. Another one of their initiatives was founding National Urban Beekeeping Day, which is celebrated annually on July 19.

The organization has since expanded and maintains 53 beehives at over 23 locations. They believe that a healthy future for our pollinators means a healthy future for all of humanity. As they support these bee populations, the bee hives are improving health, safety, and civic engagement in their local areas. Sounds like a win-win!

SAVING THE BEES

The Netherlands is showing the world how to save the bees!

Bees are a vital part of a healthy environment. As pollinators, they help plants and trees grow and thrive, which in turn, supports other species to grow and thrive. They also help us in a big way by pollinating more than 70 of the 100 crop varieties we depend on for the global food supply.

Unfortunately, bees are under threat. Pollution, pesticides, and habitat loss are putting these important insects at risk. That's why the Netherlands, which is home to around 360 bee species, is taking action.

In 2018, the country launched its National Pollinator Strategy, which consists of 120 initiatives aimed at supporting bees. With things like more wildflower meadows, insect hotels, and nature-inclusive farming, the strategy aims to support bees with better access to nesting sites and food.

By 2030, the Netherlands hopes that bee populations will be stable or at least demonstrating a positive increase in numbers. By 2023, it hopes to both reduce the number of species on a downward trend and increase those on an upward trend, each by 30%.

BUILD FOR BEES

A college student in Tennessee launched a nonprofit to help make saving the bees more accessible and fun!

In 2018, high school junior Emily Huffstetler was alarmed by a one-third loss of American honey bee colonies. Emily thought she might be able to help. Over the past 5 years, she learned a lot about raising mason bees, a species that's 90% more productive at pollination than honey bees.

She developed a Girl Scout Gold Award project called *Build for Bees*, where she taught people at local schools, nursing homes, and festivals to build houses for mason bees out of recycled materials.

Upon completion of her Gold Award project, Emily committed to ramping up her efforts. As a first-year university student, Emily secured $10,500 in grant funding and transformed Build for Bees into a nonprofit organization by 2021.

Build for Bees makes saving the bees accessible, sustainable, and fun. The organization provides free opportunities to learn about bees, including workshops, videos, and printouts. It installs solitary bee houses in community gardens and orchards, as well as region-specific bee habitat restoration kits.

STEPS YOU CAN TAKE

visit sambentley.co.uk/gnpe/city to access relevant links

○ Plant a variety of native and bee-friendly flowers in your garden. Flowers help feed bees and provide great sources of pollen and nectar. Research the best native bee-friendly flowers for where you live, as that's what bees in your area were designed to pollinate. If you're in the United States, you can download the Lawn To Wildflowers app to help you find flowers for pollinators in your state.

○ If you have to mow your lawn, do so less frequently. In the springtime, let bees feast on clovers and dandelions, until other flowers have bloomed. Earn bonus points for transforming your lawn into a mini meadow, and let it grow with an abundance of native wildflowers!

○ Set up a bee home in your garden! Now more than ever, bees need safe spaces to recover and repopulate. You can buy easy to set up bee homes online or even build your own home and create a unique space for bees. You could get started by following Build for Bees on Instagram (@buildforbees), and there's a tutorial on their website to make a home for bees out of an unwanted mug and some paper straws.

○ Leave out a dish of water full of corks or stones for bees to drink from. There's a reason being a busy bee is a phrase. Those critters work hard. And like us, bees get thirsty after working all day. Filling up a shallow dish with water can help keep them hydrated and does wonders for our planet. Be sure to add cork or twigs for the bees to land on so they don't drown!

○ Buy local and organic produce when shopping. Pesticides do serious damage to the bee population. They impair a bee's ability to navigate and reproduce and also weaken its immune system. Supporting local organic farmers who don't use pesticides is a great step to take to help save the bees. Plus, buying local produce is better for the environment, too!

CHAPTER 17

Flood Prevention

Floods are a huge problem, and they're getting worse. Floods are among Earth's most common—and most destructive—natural hazards.

Floods affected more than two billion people worldwide between 1998 and 2017. The United Nations says floods affect, on average, approximately 250 million people around the world each year and cause more than $40 billion in damage worldwide annually.

Flooding occurs when water overwhelms normally dry land, which happens in a multitude of ways across the world.

The most common flood types are:

- **Flash floods** are caused by rapid, unexpected, and excessive rainfall that raises water heights quickly.
- **River floods** are caused when consistent rain or snowmelt forces a river to exceed capacity.
- **Coastal floods** occur in low-lying coastal lands caused by storm surges, tropical cyclones, tsunamis, and sea-level rises.
- **Groundwater floods** occur when prolonged periods of rainfall causes underground water level rises to reach the surface.
- **Sewer floods** are caused by rainfall overwhelming sewage systems or treatment plants, causing the release of untreated raw sewage, bacteria, and chemicals, which has both serious health and environmental implications.

More extreme weather from climate change means floods are only increasing in frequency and intensity. The increase in temperatures accompanying global warming can contribute to hurricanes and melting glaciers contributing to a rise in sea levels. These contexts have created long-term, chronic flooding risks for places from Venice to Florida to the Marshall Islands.

Not only are floods increasingly common, but they are also causing more damage and are taking longer to recover from than ever before.

Communities most vulnerable to floods live along coastlines or within floodplains. Lack of warning systems or weather-resistant buildings can also increase risk of harm. We must do whatever we can to protect people and places from floods before it's too late!

COPENHAGEN SKATE PARK

What's better than a skate park? A flood-resilient skate park!

The Høje-Taastrup area near the capital city of Copenhagen, Denmark took flood resilience seriously when planning to transform itself into a more attractive, livable, and climate-resilient community. One way they did this was by creating an area-wide drainage system that can handle heavy rains but also acts as a skate park and recreation space for the community year-round.

The rainwater drainage system for the area is disguised as one of the world's longest skate parks, including rain-absorbent and beautiful recreation spaces. Rainwater wends its way through rain gardens until it reaches the park and the open rainwater pond, where an irrigation system collects rainwater to irrigate the park. Surplus rainwater from heavy rainstorms travels from the pond into the skate park. The skate park also acts as a detention pond in case of future community water management needs.

To increase the level of sustainability of this project, materials from previously developed infrastructure were reused—decreasing its carbon footprint and saving resources.

Approximately 5,000 tons of concrete from old bridges were crushed and reused as road base, and dredged soil was integrated back into the park's green areas.

PAKISTAN'S MANGROVE SUCCESS

Pakistan is setting the standard for successful mangrove conservation!

Pakistan has set a precedent for mangrove forest rehabilitation and should be regarded as an inspirational blueprint for other nations seeking to revive these valuable ecosystems.

Mangroves are often disregarded as mosquito-ridden thickets to be avoided, but they actually provide many benefits to the coastlines where they're found.

They are home to many important species of birds, fish, and mammals. They protect coastlines from erosion and minimize flood risk during cyclone season. And perhaps most notable in our current environment is that they are huge carbon stores, sequestering up to four times as much carbon as terrestrial forests, and they cleanse water sources by filtering out sediment.

More than a third of global mangrove populations have been depleted since 1980. Many countries have implemented regeneration plans with varying success rates, but Pakistan has been the most successful to date with an 80% success rate in their regeneration growth.

Implementing policies that give official legal protection to mangroves and creating local involvement has been vital to the success of these programs.

This case study speaks to the importance of taking the time to fully understand the organisms you're working with and gaining local support in order to foster the most successful outcome. The future of our environment looks hopeful with the help of these mangrove-protection policies.

STEPS YOU CAN TAKE

visit sambentley.co.uk/gnpe/city to access relevant links

O Get into rainwater harvesting by collecting rainwater to reuse and reduce your need for water from your main water supply. It also helps reduce bills, too! Typical households reuse rainwater to irrigate gardens, flush toilets, wash cars, and do laundry. Treehugger has a really helpful how-to on their website: A Beginner's Guide to Rainwater Harvesting.

O Build a green roof or rooftop garden to make your house or apartment building more rainwater absorbent. Portland, Oregon offers a Do-It-Yourself Ecoroof Guide which is based on ecoroof use and research since 1996.

O Use climate-native grass and plant species in your garden. Make as many natural water-permeable areas as possible to collect and filter rainwater, like rain gardens and ponds. Don't know which plant species are native to your area? Here are some free plant-identification apps to get you started: PlantSnap, iNaturalist, and Pl@ntNet.

O Reach out to your local government to invest in climate-adaptation strategies or nonstructural policy measures, like reducing development in flood-prone areas. Also ask for an improved flood warning system! Learn more about how to reduce flood risk in your city on the C40 Knowledge Hub.

CHAPTER 18

Greenifying Public Spaces

During the global COVID-19 pandemic, many of us found solace in our gardens, local parks, and even in the small spaces of greenery outside our homes. When real life came to a grinding halt, nature remained steady and comforting amidst hard times.

The pandemic taught us just how important green areas in our communities are to our mental health, and now projects are popping up all over the world to make greenifying public spaces a priority.

The pleasing aesthetic and mental health improvements that come from greenifying these spaces are just a few of the benefits. Greenifying our cities and towns also gives us a chance to tackle air pollution, fight climate change, and protect biodiversity, as well as serving to help mitigate against things like flooding. In fact, one study has revealed trees can help to cool land surfaces by up to 22°F (12°C). This is a huge win against a warming world!

As the world's population continues to grow, making public areas greener will be key to becoming more sustainable. And in some countries, it's already being done with amazing results. For example, when Singapore implemented a Greenery Skyrise Incentive in 2009, green spaces rocketed from 36% across the city to 47%, and the city is now on its way to becoming the greenest city in the world.

Fortunately, greenifying public spaces doesn't have to start at the government level. In fact, it can start right on your doorstep—all you need to do is take action.

LIVING PILLARS

Plants are being grown on lampposts in London to help clean city air!

In congested urban spaces, there can be very limited options for adding more greenery. You can't just carve out holes in the pavement to plant trees. But what you can do is use the existing street furniture that's there! This is where the Living Pillar comes in.

The Living Pillar, developed by Scotscape, uses the existing resource of street furniture (including lampposts) as a framework on which to grow plants, bringing the benefits of urban greening to increasingly busy cityscapes.

Vegetation in urban landscapes brings many benefits: from helping clean the air, expanding urban biodiversity, reducing noise, to promoting psychological well-being for local people, too!

Who wouldn't love to walk down a street lined with flowers blossoming and bees buzzing around?

BEE BUS STOPS

Leicester, England has turned its bus stops into gardens for bees!

As part of a major program to replace and revamp the city's bus stops, the Leicester City Council created a citywide network of plant-topped "living roof" bus shelters to help support local pollinators.

The bee bus stops—which are part of Leicester's ongoing Bee Roads program—are planted with a mix of wildflowers and sedum plants, a favorite among pollinating insects. These additions to the city will help support biodiversity, absorb rainwater falling on the roof, and help regulate air pollution.

Another cool part of the program is that it's created around 3.5 miles (3.6 kilometers) of roadsides and roundabouts across the city. These areas have been planted with wildflowers to help make the city a haven for pollinators by providing food stops for them to refuel on their way to larger parks and nature reserves.

Supporting local biodiversity is just one of the many goals in Leicester's Climate Emergency Strategy, which has set out an ambitious vision for the city to become carbon neutral and adapt to the effects of global warming by 2030 or sooner.

MADRID'S WIND GARDEN

Madrid is building a wind garden to help cool down the city!

The spiral design of El Jardín del Viento (Garden of Wind) within the Parque Central of Madrid will catch cool breezes above the trees and draw them down to cool the streets.

With temperatures rising all over the world amidst the global warming crisis, cities like Madrid have been thinking outside the box for ways to mitigate the negative effects this warming causes. Inspired by ancient Middle Eastern wind towers, the wind gardens could help cool the city by an estimated 7°F (4°C).

The wind towers are designed by Dutch landscape architecture firm, West 8. Their spiral structures are lined with ferns and mosses that operate above the canopy level of other trees to catch breezes high in the sky and funnel them down to street level to cool the city.

This invention, along with other innovations, will help make hot city living more enjoyable and allow people to remain in the city even in the height of summer, which could reduce travel emissions and support the local economy.

With these towers in place, there will be less need for air conditioning homes, which will reduce energy consumption for all. Other cities have started similar projects, like in Tulum, Mexico, where a new train station will be passively cooled by funneling in the sea breeze, which reduces the need for artificial ventilation. Meanwhile in Athens, the city is turning abandoned plots of land into "pocket parks" to reap the cooling benefits of increased vegetation cover in cities. The future is looking cooler than ever with these innovations!

RECONNECTING WITH NATURE

Dutch volunteers transported 1,000 trees in wooden containers to create a forest in the city of Leeuwarden to help local people reconnect with nature!

An initiative started by a landscape architect in the Dutch city of Leeuwarden has taken off as the idea of a walking forest has transformed the urban landscape.

Initially created in May 2022 as part of Arcadia, a triennial art festival based around reconnecting people with nature, the trees have become so popular that people are keen for them to stick around.

The trees were moved around the city to simulate what a brighter future could look like if urban planners adopted a greener mindset. People of all backgrounds have become invested in the trees and readily volunteer to move them around the city.

Each tree is equipped with a QR code that anyone can scan to learn about the 60 to 70 native tree species that are part of the walking forest. This is a great way to gain local interest in the project and increase support.

The trees offer shade and create a more relaxing environment, which is beneficial to the mental health of city dwellers, while also having a huge impact on minimizing the urban heat-island effect.

A hundred days after the trees arrived, they were permanently planted around the city, particularly in lower income areas where there was less greenery.

This is hopefully a promising step in helping Friesland to achieve their goal of becoming the most circular region in the European Union by 2025 and sets an example that other cities can follow.

STEPS YOU CAN TAKE

visit sambentley.co.uk/gnpe/city to access relevant links

O Head out for a walk in your neighborhood. Are there areas where space can be utilized to plant trees and grow wildflowers—or can an abandoned car park be turned into a local garden? Think big or small, and reach out to the wider community and decision makers to see what's possible!

O Check out the Ron Finley Project, which is all about teaching communities how to transform food deserts into food sanctuaries. It all started when Ron realized his local area in LA struggled to offer fresh produce to residents. He began guerrilla gardening by planting vegetables in dirt patches next to streets and inadvertently kick-started a revolution.

O Find your local community groups and see what environmental initiatives you can get involved in. Whether it's joining a community-shared allotment group or helping to raise awareness with education and petitions, there's always something to jump into.

ENERGY

Renewable energy technologies will help break our reliance on fossil fuels, boost national energy security, and help us fight climate change.

We're running out of fossil fuels. There are only about 50 years of gas and oil left, with 114 years of coal left, based on current reserves. But we can't afford to wait for fossil fuel reserves to run dry before implementing serious change.

Burning fossil fuels releases the greenhouse gas carbon dioxide, which is one of the largest drivers of man-made climate change.

Climate change is already causing heat waves, floods, and other extreme weather events across the globe. Other gasses produced by burning fossil fuels, like in car engines, contribute to local air pollution, which is linked to 7 million premature deaths annually.

So what do we do? The first and most important thing is to reduce how much energy we use individually and more importantly as a society. Despite an increased awareness of the impact of energy generation on our climate and health, global consumption of energy is rising.

There is no alternative. We must learn to reduce our demand, particularly as the richest 10% of the global population—people with net income over $38,000— are responsible for 52% of greenhouse gas emissions.

We all need to use some energy to live (and that's okay!), so once we've cut demand and reduced waste, what's the next step? We need to start generating our energy in ways that don't produce carbon. This is called the *decarbonization* of the energy sector.

We have the technologies to do this, including renewable energy technologies and capturing energy from nuclear fission, which comes from the decay of radioactive atoms. Given our growing energy demand, the question is not which low-carbon energy technology we should use, we need to use all of them!

Renewable energy is energy generated from sources not depleted when used. These energy sources are all around us in the natural world. We can capture energy from fast-flowing water, from the tides, from the wind, and from the sun. We can capture heat from the sun, the earth, and the air, using ground-source or air-source heat pumps.

These technologies can supply the energy demands of homes, businesses, and with the right infrastructure, the whole of a community or country.

Despite the need to use some fossil fuels to produce these technologies, capturing energy from renewable sources releases significantly less greenhouse gasses into the atmosphere than fossil fuels across the lifetime of the technology. Put simply, we need more energy produced by solar, tidal, wind, and nuclear technologies!

CHAPTER 19

Solar Power

Every 2 hours, the Earth's surface is bathed in sunlight containing more energy than humanity uses in a year. This staggering figure shows the true potential of solar energy for innovation.

The idea of capturing solar energy is not new—plants capture the sun's energy and convert it into sugars for growth. By eating plants and burning wood and biomass, we make use of this captured energy.

However, harvesting plants for fuel also causes biodiversity loss and damages the environment, and burning biomass causes air pollution that can damage people's health. Fortunately, solar energy can be captured directly and efficiently using man-made technologies. There are two commonly used technologies for capturing solar energy.

Photovoltaic cells, more commonly known as *solar panels* or *PV panels*, convert sunlight into electrical energy. These can be easily installed and supply electricity directly to street lights, bus shelters, houses, sheds, and larger-scale installations. Any excess electricity can be sold to the national grid, covering and potentially even exceeding installation costs.

The sun can also meet hot water demands via solar-thermal panels. Cold water passes through the panels, the water is heated by the sun, and is stored in a hot water tank.

Renewable technologies still have a carbon footprint because of resources used to make them, but the footprint is dramatically less than fossil fuels. Excitingly, the uses of sunshine go beyond solar farms and solar panels on roofs.

SOLAR POWERING EUROPE

There are now enough solar installations to power all of Europe!

The world hit a major milestone in 2022 as we saw enough solar panel installations to generate 1 terawatt (TW) of electricity—enough to power the whole of Europe!

While still just a fraction of global energy supply, this figure is technically enough to meet the electricity needs of nearly every country in Europe combined (around 3,050TW hours).

Distribution and storage limitations means that, at the moment, solar only makes up a small part of Europe's energy use. The European Union meets 3.6% of its electricity demands from solar power, with the United Kingdom at 4.1%.

However, with record growth seen in 2021 for residential installations, China, Europe, and the United States are driving the growth of solar energy harvesting—with solar electricity generation becoming more energy efficient and cost-effective than ever before.

SOLAR PANELS AT NIGHT

Solar panels that can generate electricity at night are sparking hope for the future!

As solar panels pick up in popularity around the world amid a global energy crisis, new innovations are key—and, fortunately, some engineers are already ahead of the game.

Most notably, a team at Stanford University has been able to develop a solar cell that can generate electricity at night, instead of just storing the energy it saves up through the day in batteries.

The result? A "continuous, renewable power source for both day and nighttime," according to the study published in the journal *Applied Physics Letters*.

The device's thermoelectric generator draws electricity from the minute temperature difference between the solar cell and ambient air.

Those behind the project say that it can provide nighttime standby lighting and power off-grid and mini-grid applications (essentially, independent energy networks where communities are smaller or too far from the main grid).

But as demand for solar installations and solar jobs rise, it's no doubt just the first step on an exciting journey ahead.

PORTUGAL SOLAR PARK

Europe's biggest floating solar park is coming to Portugal!

Around the world, countries are setting up floating solar-panel parks in a bid to accelerate the move away from harmful fossil fuels.

In early 2022, Portugal set up Europe's largest floating solar farm in Alqueva in the south of the country. The Alentejo region is home to Alqueva Lake, which is the largest artificial lake in Europe.

The new solar park is home to nearly 12,000 panels (that's about four soccer fields' worth). Built by EDP, the largest energy producer in Portugal, it has the capacity to supply 30% of the surrounding population with power.

The new project is part of EDP's aim to "go green" before the end of the decade. It has pledged to only produce energy via wind, sun, or water. The company has already abandoned coal production completely in Portugal.

Floating solar parks can also be found in countries like China, Singapore, Thailand, and the United States. The world's largest is found in Dezhou, China.

MANDATORY SOLAR

Solar panels on rooftops could make Europe greener than ever!

Homes across the European Union could become greener than ever under new proposals for mandatory solar panels on all new buildings (and it couldn't have come at a better time).

It's all part of a "solar rooftop initiative" launched by the European Commission that aims to scale up and speed up renewable energy in power generation, industry, buildings, and transportation.

Not only will the proposal see a move away from dependency on Russia for fossil fuels by 2027, it'll also boost the continent's green transition and reduce energy prices over time, which is very welcome news as figures continue to climb.

It's not just about rooftop solar panels, though. Among the many objectives within the proposal, which is ambitiously working to hit even better renewable energy targets than planned by 2030, they'll be doubling wind and solar capacity while encouraging member states, businesses, and citizens to reduce energy consumption on a daily basis, too.

After all, from using more public transportation to turning off lights, it's the energy we don't use that saves the most.

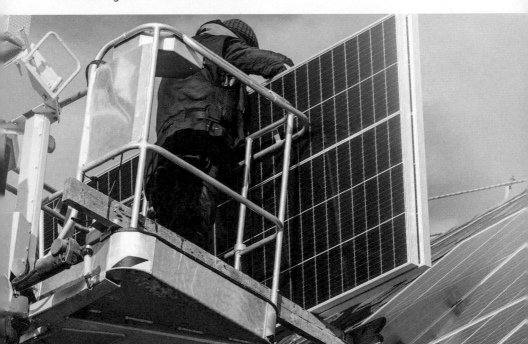

ACCESSIBLE SOLAR

Delaware is giving free solar panels to low-income residents to help make green energy more accessible!

Green energy is great for the planet, but for individuals, it can be costly to set up. That's why Delaware's Department of Natural Resources and Environmental Control (DNREC) recently launched a new initiative for 50 households every year, called the Low-to-Moderate-Income Solar Pilot Program.

DNREC's new program disregards low credit scores and offers low-income households totally free solar panel installation, up to 4kW.

It also helps out moderate-income households by paying 70% of the installation costs, up to 6kW. The household then just has to pay the remaining 30%.

The new initiative not only allows lower-income families and individuals access to cleaner, more sustainable power, but it will also help reduce household expenses in the long run.

Low-income households (classified as those who bring in $46,060 or less) spend a significant percentage of their income (around 30%) powering their homes. Across the US, more than five million households face energy poverty.

COVERING CANALS WITH SOLAR PANELS

To offset widespread drought, California is covering canals with solar arrays!

A new pilot program will see some of California's irrigation canals (artificial waterways that carry water to farms) covered with solar panels in a bid to save more water and boost access to renewable energy.

When solar panels are placed over canals, they help prevent water from evaporating. In fact, if all of the state's irrigation canals were covered, more than 63 billion gallons (238 billion liters) of water could be saved.

Inspired by a similar trial in India, Project Nexus, created by Turlock Water & Power, will first install solar panels in two locations: Hickman and Ceres.

As well as saving water, the panels could also remove the need for diesel generators, which are usually used to pump water along the canals.

Even better, if the project is successful and solar panels end up covering all of California's irrigation canals, they could produce enough power for around half of the homes in Los Angeles.

SMARTFLOWER

This huge sunflower is made from solar panels!

It's called a *SmartFlower*, and like a sunflower, it tracks the sun throughout the day, producing clean, renewable energy with 12 petals made from solar panels.

Each morning, SmartFlower automatically unfolds when the sun rises and follows the sun at the perfect angle throughout the day. In the evening when the sun sets, it even packs itself away for the night.

The solar modules also clean themselves when opening and closing, which helps make SmartFlower even more efficient.

If there's a storm or dangerous weather, it automatically folds into a secure position, too. It also has a simple installation process to ensure it's easy to move and can be installed in just 2 to 3 hours.

The SmartFlower can be used by businesses or in your own home. It typically produces anywhere from 4,000 to 6,200kWh per year. Wouldn't you love to have one of these?

STEPS YOU CAN TAKE

visit sambentley.co.uk/gnpe/energy to access relevant links

○ Install solar PV panels at your property. You'll need to find out if any form of permission is required before starting, and what the best way to go about it is. If you're in the UK, you can visit the Energy Saving Trust website, which also has a solar energy calculator where you can get a better idea of the benefits you may see from installing a solar PV system. If you're not in the UK, search online for *how to install solar PVs in my area*.

○ Check out DIY Solar Power with Will Prowse on YouTube (@willprowse). He has a bunch of beginner-friendly videos about solar power, from videos on how to build your first solar power system to how to size up your solar power system.

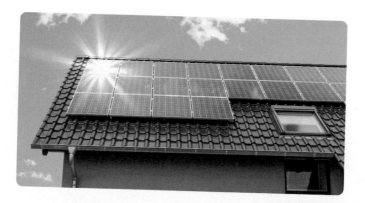

CHAPTER 20

Tidal Power

Our oceans' tides are one of the most powerful natural forces on the planet, and we can harness them to generate electricity. According to the US Energy Information Administration, waves off the coast of the United States could generate around 2.64 trillion kWh of electricity per year.

Tides are caused by the gravitational pull of the moon—and to a lesser degree, the sun—on the oceans as the Earth rotates. The steady rise and fall of our oceans provide a regular, predictable source of clean tidal energy.

There are three main ways to capture tidal energy: tidal streams, tidal barrages, and tidal lagoons. Tidal stream energy technologies place a turbine, similar to a wind turbine, into the path of flowing tidal water. Water is much more dense than air, so tidal energy is more powerful than wind energy and the turbines can turn more slowly.

Tidal barrages are dam-like structures built across tidal rivers, bays, or estuaries, blocking receding waters during high tide and releasing it at a controlled rate during low tide to drive turbines that generate electricity.

Tidal lagoons are bodies of water partially enclosed by natural or man-made barriers, submerged at high tide. They can be built along natural coastlines. Tidal lagoons can generate near continuous power because they're able to generate electricity when filling and emptying.

Tidal technology is promising but has not been developed as much as other renewable energy sources due to high costs, which will hopefully be made more affordable soon!

WAVE ENERGY CONVERTER

A Swedish company has developed a wave energy converter that can turn ocean waves into electricity!

Eco Wave Power creates clean energy by installing its system on existing man-made structures, such as breakwaters, piers, and jetties.

How does it work? The floaters pull energy from the rising and falling motion of the incoming waves, which is then transferred to the grid as clean energy.

As the floater moves up and down with the waves, it drives a hydraulic piston that compresses fluid that's then stored in an accumulator.

When the fluid is released, it rotates a hydraulic motor that powers an electric generator, producing clean electricity.

At a former World War II ammunition jetty in Gibraltar, Eco Wave Power installed a 100kW system that has produced power for the grid in Gibraltar, powering around 100 homes over a 6-year time period. It's said to be the only grid-connected wave energy array in the world.

More recently, plans were announced to relocate the energy conversion unit from Gibraltar to a 35-acre (14-hectare) campus located at the Port of Los Angeles to bring its pioneering wave energy to the United States. Here it has the potential to produce 15MW of electricity, enough to power around 15,000 households.

SEA WAVE ENERGY

A device has been created that can float on top of waves and convert wave energy into electricity!

Sea Wave Energy Ltd. (SWEL) has created this amazing device called the *Waveline Magnet*, a unique wave energy converter (WEC).

The Waveline Magnet has been in development for over a decade and has been tested in multiple real-sea trials. It has also been tested in state-of-the-art wave tank facilities, where the system was put to the test through harsh sea conditions. The system worked perfectly throughout the whole process.

So how does it work? The Waveline Magnet is made of lots of floating platforms linked by a spine-like power system. As a wave passes through the system, it seamlessly follows the wave movement, generating power as the wave rises and falls.

The power production capabilities of the Waveline Magnet means it can be used in a bunch of different ways, from water desalination, electricity production, and even marine farming.

The low-mass materials used, like plastics and reinforced plastics, lower the manufacturing cost, making the Waveline Magnet easy to repair and maintain.

SWEL believes their technology can compete with nonrenewable sources, stating that "SWEL has developed and tested their devices in wave tanks and live sea environments. The R&D (research and development) indicates that even in its current developmental state, the Waveline Magnet can produce substantial power levels at an exceptionally low cost, competing even with nonrenewable sources."

In addition to this, the Waveline Magnet can be made from durable recyclable materials, meaning it's the only wave energy converter that can be 100% recycled, according to SWEL.

Next in SWEL's journey to shake up the energy industry is to design their first commercial system, find strategic partners to help take it to the next level, and carry out final tests before taking it to market.

WAVE ENERGY RESEARCH

The US Department of Energy (DOE) announced $25 million in funding to support cutting-edge wave energy research!

This funding went toward supporting eight projects that are making waves in the clean energy sector. The eight projects focus on three sectors:

- Testing wave energy converter designs in remote regions or on small-scale energy grids.

- Developing wave energy converter designs to connect or disconnect from the electricity grid.

- Researching and developing environmental monitoring technologies, instrumentation systems for controlling wave energy converters, and other technologies.

The biggest funding recipient, CalWave Power Technologies, Inc. received $7.5 million to help them create technology that operates entirely below the water's surface and moves with the motion of the waves, enabling it to produce electricity. They successfully deployed a small-scale version of this technology out at sea for 10 months, but this funding will allow them to scale up their ideas with the aim of deploying their full-scale devices at sea, creating clean energy over a 20-year period.

"Harnessing the unrelenting power of the ocean is a clean, innovative, and sustainable way to curtail carbon pollution—benefitting American businesses and families, especially coastal communities hit hardest by the impact of climate change," said US Secretary of Energy Jennifer M. Granholm.

The hope is that these technologies can help decarbonize the grid and reach the United States' goal of net-zero carbon emissions by 2050.

STEP YOU CAN TAKE

visit sambentley.co.uk/gnpe/energy to access relevant links

○ Keep up-to-date with new wave technologies being tested out at PacWave, a grid-connected, full-scale test facility—the first of its kind in the United States—funded by the Energy Department. Discover the latest news on the PacWave website.

CHAPTER 21

Wind Turbines

Wind is the power of our ancestors. Humans have been harnessing wind power for many millennia—sailing boats to provide fast transportation across oceans, windmills to pump water or grind flour, and even wind machines to make music.

Wind turbines that use wind to generate electricity were invented in the late 1800s, even before electricity was commonplace. Unfortunately, the adoption of wind power has been hampered by competition with fossil fuel-based energy generation. However with increasing fuel prices and the harmful effects of burning fossil fuels, wind power will again become increasingly important to fulfill our energy needs.

The two main types of wind turbines are horizontal axis and vertical axis. Horizontal axis wind turbines are the more effective kind, so they're more frequently used for large scale energy generation (e.g. wind farms). Vertical access wind turbines enable wind activation from any direction and are often used to power individual properties.

Although wind turbines can kill birds that collide with them, innovations to reduce bird deaths are being introduced. When considering this problem, it's worth remembering that other power generation methods, climate change and even cats, kill many more birds than wind turbines. Innovations and environmental wins in the wind energy space are becoming more and more frequent, so let's take a look at some news that doesn't blow!

RECYCLABLE WIND TURBINE BLADES

The first 100% recyclable wind turbine blade can be reconstructed over its lifetime!

One challenge facing the wind energy sector has been how to bring it into the circular economy. The good news? We may be about to see production of the first-ever recyclable wind turbine blade.

Launched in September 2020, the Zero wastE Blade ReseArch (ZEBRA) project is a culmination of work from some of the world's most forward-thinking companies in ecodesign, such as Arkema, CANOE, Engie, LM Wind Power, Owens Corning, and SUEZ. Led by French research center IRT Jules Verne, this project is a pioneering partnership in the energy sector.

Now, they've created the first ever prototype of a 100% recyclable wind turbine using Arkema's Elium resin (a thermoplastic well-known for its recyclable properties), which means the blade can be reconstructed as it ages.

Today, the 203ft (62m) blade is being put through a full-scale structural lifetime testing with LM Wind Power to ensure it hits all the key requirements for commercial production.

Expected to finish in 2023, it's believed that by the end of the project, the teams will have overcome the obstacle of introducing the wind energy sector into the circular economy loop in a viable way.

ZEBRA 100% Recyclable

TULIP TURBINES

Have you ever seen tulip-shaped wind turbines before?

These turbines, created by Flower Turbines, produce clean energy at low-wind speeds from any direction.

Due to their shape they're also safe for birds; they create no more noise than the wind; and when grouped together, the Flower Turbines actually drive wind into one another, meaning they can perform up to 50% better. They even have the potential to produce more power per square meter than solar in windy areas!

They come in a variety of colors and sizes. There's a 3ft (1m) model that is ideal for off-grid projects, a 10ft (3m) model that is their most popular and can produce one-third of the energy consumption of a household, and a 20ft (6m) model for large-scale production of electricity.

So from helping to charge e-bikes to powering office blocks, the Flower Turbines can cover a wide range of uses!

UPCYCLING WIND TURBINES

Old wind turbine blades are finding new life as bridges!

What do you do with a wind turbine that's no longer able to generate energy? Well up until now, the difficult answer was simply sending them to the landfill (or off to the incinerator). Thanks to some innovative minds, though, they're now finding new homes as bridges.

The concept started in Poland and has now reached Irish shores, with the second-ever bridge made of recycled wind turbine blades built in County Cork in 2022.

The engineers and entrepreneurs behind the bridges hope they're the start of a new trend: repurposing old wind turbine blades for infrastructure projects that would otherwise take up copious amounts of space in landfills.

Even better, this project also saves in the energy required to make new construction materials, since the blades are made of sturdy materials that don't break down easily and still have decades of life in them.

In fact, the developers behind the Polish project reckon these new bridges could last for "at least a hundred years," making this a no-brainer in the future of bridge construction.

OFFSHORE WIND FARMS

Italy has established its first-ever offshore wind farm!

The new farm is located off the coast of Taranto, the third-largest continental city in southern Italy. It's home to 10 wind turbines and is also the first of its kind in the Mediterranean Sea. It has the capacity to produce enough power for roughly 60,000 people.

Over the course of 25 years, it could also save 730,000 tons of carbon dioxide from entering the atmosphere.

In a bid to accelerate a move away from fossil fuels and a reliance on gas produced by Russia, Italy is now embracing wind farms. In 2022, the government approved another six wind farms, although these will be built on land.

In a bid to encourage more companies to set up offshore wind farms, the Italian government announced in 2022 that it would be offering incentives and grants. It has also made it far easier for developers to apply to build them.

WIND FARM JOBS FOR WALES

Floating wind farms in Wales will help create 29,000 jobs!

In the United Kingdom, offshore wind turbines are usually fixed into land or the seabed. However, new plans could see more floating wind farms (which can be set up in deeper water and in areas where there is more forceful wind) appear off the coast of Wales and Cornwall in the Celtic Sea.

While some are against wind farms for allegedly disrupting wildlife on land and spoiling the look of the countryside, floating wind farms do not present this problem, as they're not visible from shore.

In 2022 the Crown Estate, a property business that is technically owned by King Charles III, but is run independently, revealed it would lease space for new floating wind farms 40 miles (64 kilometers) off the coast of southern Wales.

The project could help to create 29,000 new jobs in total and potentially produce power for up to four million households by 2035. The turbines will likely be built in Port Talbot and Pembrokeshire before they are taken out to sea.

STEP YOU CAN TAKE

visit sambentley.co.uk/gnpe/energy to access relevant links

O Look at the possibility of setting up a domestic wind
turbine for your home. One option mentioned are the
Flower Turbines, which you can learn more about at on
their website. You need to look into doing a detailed,
site-specific assessment to see how effective a
microwind turbine would be for you and any
permissions you need to get. Start off by searching
how to install a domestic wind turbine for my home.

CHAPTER 22

Energy Storage

The sun doesn't always shine, the wind doesn't always blow, and the tides work on their own schedule. Our energy demands are also constantly varying, in predictable and unpredictable ways.

Relying on renewable energy sources without saving some of that energy for later use will result in disruptions and power cuts. New and existing energy storage technologies store electricity for use later or transport it to wherever it's needed. Cheap, efficient, and environmentally friendly energy storage solutions will be essential to turn variable renewable energy sources into a constant supply of clean electricity for the long haul.

So what about batteries? When thinking about energy storage, most of us reach for batteries, which store electricity in the chemicals inside. However, batteries currently make up only a small proportion of energy storage technology for the power grid, as they are better for short-term rather than long-term energy storage. They are relatively bulky and expensive and are often made of rare metals.

For example, lithium, the major component of batteries in our phones, laptops, and electric cars, is predicted to hit worldwide shortages between 2025 and 2050, depending on demand. Researchers are working on ways to recycle rare metals used in batteries and are developing cheaper, smaller batteries that use fewer or no rare materials.

Over 90% of global grid-power storage capacity is pumped hydropower, which stores electricity as gravitational energy. How does this work? Surplus electricity is used to pump water to the top of a hill. When the energy is needed, the water is allowed to flow back down, turning turbines to produce electricity as it goes. Varying the flow varies the amount of electricity produced, allowing the plant to be responsive to energy demands, which fossil fuel and nuclear power plants are less able to do. Energy storage is much cheaper and longer-lasting than batteries. This technology requires major infrastructure, so this can have a significant environmental impact.

Let's quickly talk about hydrogen and other e-fuel. E-fuel is a group of zero- and low-carbon chemicals that can replace fossil fuel as an energy source, reducing dependence on global trends in oil and gas prices, while also reducing pollution. Electricity (from renewable technologies) is used to produce e-fuel. The energy is stored in chemical bonds that can be broken down (by burning) to reclaim the energy. The e-fuel can be more easily and cheaply stored and transported than the electricity.

Hydrogen is an e-fuel with a crucial role in achieving the net-zero targets of many countries. It can be created in a zero-carbon way using electricity to split water into hydrogen and oxygen or a low-carbon way from methane and steam, capturing the carbon-containing pollutants produced before they reach the atmosphere. Once created, the hydrogen can be used by to make fertilizers, to reclaim the electricity in a fuel cell, to power vehicles, or heat your home. Up to a fifth of natural gas currently used in UK homes to heat and cook could be replaced by hydrogen as early as 2023.

Hydrogen can also be combined with carbon dioxide to make other e-fuel with different properties, using electricity. E-methane is one example, while e-petrol and e-diesel are still being researched. Various e-fuels have different properties, so it's useful to have a range for different purposes.

SAND BATTERY

Could sand batteries solve green energy's big storage problem?

There's always been one key problem with green energy, and that's how to store it. Luckily, the problem could be on the cusp of being solved. How? Finnish researchers installed the world's first fully functioning sand battery, which stores clean energy for months.

Essentially, it means we could potentially solve the problem of year-round supply with a simple, cost-effective way of storing power for when it's needed most.

At the moment, green energy can only be stored in batteries made with lithium, which are expensive and have a large physical footprint (and can only cope with a limited amount of excess power). With sand batteries, energy is stored as heat at around 932°F (500°C), meaning homes could be warmed through winter, when energy is more expensive, at a cheaper rate.

Now, it's hoped the technology can be scaled up to make a big difference, so that, one day, it will be able to efficiently return electricity to the grid.

In the meantime, it could be just what's needed to replace heat generated from fossil fuels for things like cooking food and creating textiles, helping us take one step closer to a green future.

MINE SHAFTS FOR GRAVITY ENERGY STORAGE

An Australian start-up is using old mine shafts to store energy, allowing renewable energy to be used when needed!

The start-up Green Gravity uses technology that raises and lowers heavy weights in a mine shaft to capture electrical energy as gravitational potential energy.

This is called *gravitational storage,* and I'll do my best to explain how it works. Essentially, when renewable energy generation is higher than the electricity demand, the extra electricity is used to move heavy weights to the surface. The energy is then stored there as gravitational potential energy until the energy is needed.

When energy is required—let's say during the evening when there is limited solar and wind energy but demand for electricity is high—the weights are allowed to fall back down due to gravity, turning the rotor shaft of an electric generator and producing electricity. In this way, the gravitational potential energy is converted to mechanical energy (by moving the rotor shaft) and then into electricity (in the generator).

Using these decommissioned mines is central to Green Gravity's plan, as they can utilize the size of the mines to store a great amount of energy. The mines are often connected to the electricity grid, meaning the electric infrastructure already in place can be reused, cutting down on further costs.

So far, Green Gravity has identified 175 mine shafts that it believes are suitable to house its technology and would require minimal work to be converted for energy-storage purposes.

STEPS YOU CAN TAKE

visit sambentley.co.uk/gnpe/energy to access relevant links

O A great channel to follow on YouTube is Undecided with Matt Ferrell (@UndecidedMF), which explores various energy storage solutions, along with how sustainable and smart technologies impact our lives.

O If you'd like to learn more about gravitational storage, the YouTube channel Just Have A Think has a great video on it called "Gravity Energy Storage. Who's Right and Who's Wrong?" It's definitely worth checking out!

We're all trying to get from point A to point B as quickly, efficiently, and safely as possible. Yet our current transportation systems have a lot of room for sustainability improvement.

For one: cars. I don't mean the Pixar flick. The automobile has shaped human development for over a century and is seen in many countries as the symbol of mobility and personal freedom.

But that image is changing in modern times. We now know the extent of the damage cars cause on the environment, none of which is good for the planet or the transportation industry in the long term. We can no longer build our lives around these fossil fuel-guzzling machines.

If you think you're just one person and choosing a different method of travel for your ride to work won't help the planet, think again. The average passenger vehicle produces about 5 tons of carbon dioxide in a year, and every gallon of burned gasoline emits 20lb (9kg) of CO_2.

It's important to be conscious of how harmful air travel can be, too. While necessary and convenient for many people, commercial aviation accounts for around 2.4% of global CO_2 emissions. Together with other gasses and the water vapor trails produced by aircraft, the industry could be responsible for around 5% of global warming.

On the other hand, sustainable public transport systems have the ability to provide a more positive social and environmental impact. These systems use more sustainable vehicles, energy sources, and infrastructure to move people and goods within cities or around the world.

Transportation and mobility impact our lives everyday and over the long term—from our environment to our health and well-being. So what options do we have, and what is happening around the world to make sustainable travel more accessible? The next few chapters will give you these answers!

CHAPTER 23

Electric Vehicles & Bikes

Electric vehicles (EVs) are trendy these days because of their environmentally friendly reputation when compared to their gas-guzzling cousins, traditional internal-combustion engine cars.

Decarbonizing the transportation sector is one key to keeping global warming below 1.8°F (1.5°C). And electric vehicles are the single most important technology available to decarbonize the transportation sector. One of the criticisms of electric vehicles is that in order to produce the electricity needed for them, fossil fuels have to be burned. This is largely dependent on where you get your electricity from, though. If you are using electricity that's been generated from solar energy, or even solar panels on your own home, then this means your EV is using 100% clean energy.

Let's say you live in an area where the electricity grid is still largely dependent on fossil fuels—like Wyoming, which is the worst state for this in the United States—it is still 28% cleaner to drive an EV than to drive a gas-burning car, even in this worst-case scenario. On the flip side, on an average day in California only 0.1% of their grid is powered by coal.

A report from the International Council of Clean Transportation looked deeper into this topic to study the lifetime greenhouse gas emissions of combustion engines and electric passenger cars in Europe, United States, China, and India. As electricity grids across the world continue to decarbonize, this gap between internal-combustion engine vehicles and battery-electric vehicles will continue to grow larger, too, making electric vehicles a much more efficient and ecofriendly alternative. EVs also have no tailpipe emissions, which contribute to local air pollution through the distribution of harmful gasses like nitrogen dioxide, which can cause damage to the human respiratory tract and increase a person's vulnerability to and the severity of respiratory infections and asthma. So electric cars are the future, right?

Electric cars represented 11% of all new, registered passenger vehicles in 2020. In 2012, 130,000 electric cars were sold worldwide. Fast forward to 2021 and that number has climbed to 6.6 million, an 8.5% share of global car sales.

But electric vehicles are by no means perfect. EV batteries are made of lithium ions and other resources. You may have seen the international attention of these resources, including reports of child labor being used in cobalt mines in the Democratic Republic of Congo, as well as abuses of indigenous rights around lithium mining in South America. However, there are many sources of cobalt in the world that don't depend on child exploitation, such as the Talvivaara mine in Finland, and similar mines are set to open over the coming years. Also, since 90% of EV batteries can be recycled, as we move forward we can source more cobalt from recycled batteries and extract less from the planet.

What about other electric vehicle types as more environmentally friendly alternatives to cars? Electric bicycles are increasingly replacing cars for medium-distance journeys, which reduces CO_2 emissions and costs. E-bikes are pedaling toward wider consumer adoption worldwide, even in the United States. Mindsets are shifting on the environment, health, and wellness, and maybe some folks are even seeking a little daily adventure on their work commute. They're quiet, a lot more affordable than cars, easier to maintain, and with more and more funding being allocated to bike paths and cycling programs, commuting with them should become much more accessible in the near future.

ELECTRIC VEHICLES REDUCING OIL DEMAND

Electric vehicles are reducing oil demand on a huge scale by up to 1.5 million barrels a day!

In 2021, estimates from the research firm BloombergNEF showed that the adoption of electric vehicles and fuel cell vehicles avoided almost 1.5 million barrels of oil being used per day. Oil use avoided by EVs has more than doubled since 2015, up from around 725,000 barrels of oil per day.

According to the report, two- and three-wheeled EVs accounted for 67% of the oil demand avoided in 2021. This is due to their rapid adoption, particularly in Asia.

Buses, with 16% of total oil demand avoided, were next; followed by passenger vehicles, the fastest growing segment, accounting for 13%. Commercial vehicles accounted for just 4% of total oil demand avoided in 2021, largely from light commercial EVs.

Let's hope we see more rapid adoption of EVs across the transportation sector worldwide so we can permanently remove our dependency on oil for transit!

FASTEST ELECTRIC VEHICLE CHARGER

The world's fastest electric vehicle chargers were installed in Norway, capable of fully charging a car in 15 minutes!

One of the most common concerns about electric vehicles is the amount of time it takes to charge them when out adventuring on the road. No one wants to be stuck at a charging station for hours, adding precious time onto a long distance trip.

Well, let me introduce you to ABB's Terra 360. The Terra 360 is the most powerful high-power charging solution in the world, delivering up to 360kW of power and a full charge in less than 15 minutes.

It can charge four vehicles at the same time, can be used to charge both passenger and commercial vehicles, and supports all EV standards.

The Terra 360 was piloted in the Oasen shopping center in Bergen and in the Norwegian ski resort of Geilo. After these pilot installations, the goal is to roll out more of these charging stations across Norway and Sweden!

E-BIKE INCENTIVE PROGRAM

France is offering money to people who trade in their car for an electric bike!

To help incentivize cleaner and more environmentally friendly methods of transportation, the French government created a program to pay people up to $4,260 (€4,000) to trade in their gas-powered car for an electric bike. For those not ready to give up their car, a $426 (€400) grant is available for a bike or e-bike!

Currently, only 3% of France's population has made the switch from cars to bikes, but the French government wants that number to rise to 9% by 2024. To help drive this transition to bikes, the government said it will also invest $266 million (€250 million) to make the city of Paris entirely bikeable, with the city's mayor promising to add over 80 miles

(129 kilometers) of bike lanes over the next 5 years.

A similar incentive program was also launched in Lithuania, which saw such great results that the government had to increase the budget of the program due to high demand.

Talking about the program in Lithuania, Austėja Jonaitytė, a spokeswoman for the Environmental Project Management Agency, said, "The initiative received a lot of attention from the population. The number of applications exceeded all expectations. For this reason, the Environment Ministry has allocated an additional $3.2 million (€3 million) from the Climate Change Programme."

With the appetite for such programs clearly there, let's hope we see other governments across the world launch similar plans soon!

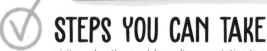

STEPS YOU CAN TAKE

visit sambentley.co.uk/gnpe/transportation to access relevant links

○ To get a deeper insight into electric cars, from how to install wall chargers to busting common EV myths, check out the YouTube channel Now You Know (@NowYouKnowChannel).

○ Carpool! By having more people using one vehicle, carpooling reduces each person's travel costs to cover fuel and tolls, and helps relieve driving stress. In 2019, 76.3% of workers in the United States commuted by driving alone, and only 9% of workers carpooled (a drop from 19.7% in 1980). Joining a carpool can help lower household fuel costs, prevent greenhouse gas emissions, and reduce traffic congestion.

○ Start car-sharing or join a car club. Avoid purchasing or maintaining cars by instead sharing the responsibility with a community by renting cars for short periods of time, even for an hour. Research which car-sharing operators are near you to get started.

○ Grab a bike when you're out and about as an alternative to car travel or public transportation. Download a bike-sharing app such as Ofo, LimeBike, or JumpBikes. Depending on your region, some of these apps may be better than others.

○ If you only need a car occasionally, you can use car-sharing apps like ZipCar that make it easy to find cars on demand. ZipCar in particular has a goal of having a fully electric car fleet by 2025, and you can specify an electric car during your selection process.

CHAPTER 24

Public Transportation

Less cars + more public transportation = more space for people to get around. Streets can take up to half of the land in cities, this means a large portion of our cities are actually habitats for cars. It's time to take back our cities from cars! Since our transportation choices are affected by urban design choices, which include transportation systems, we must demand our governments make extensive, accessible, and sustainable public transportation a priority.

Public transportation comes in many forms, including buses, underground and aboveground rail systems, ferries, and cable cars, but all are essentially systems that move people from one area to another in an efficient, affordable manner. By providing alternatives to car use, public transportation can decrease the number of private vehicles on our roads. It eases traffic congestion, benefits the environment, and is more cost effective, reliable, and accessible than road networks. It can also transform the way our cities look and how we live in them.

Cities like Portland (Oregon), Seoul, and Madrid have transformed their streets to be used more by people and improve accessibility to plants and natural spaces. These green areas replace asphalt and roads that are bad for flooding, the urban heat-island effect, and so much more.

These illustrations depict how personal vehicles take up so much more road space than public transportation or cycling does.

STOCKHOLM FERRY

A flying electric ferry will be launching in Stockholm that will be able to travel 50mi (80km) on a single charge and will be the fastest ferry in the world!

Swedish company Candela is revolutionizing waterborne transport with their latest innovation, P-12, the world's fastest electric passenger vessel. Designed as an alternative to land-based public transport such as cars or buses, Candela hopes to help reduce congestion and pollution by offering an option that is quicker than trains, buses, or cars during rush hour.

Thanks to hydrofoils—fins that lift it out of the water—the P-12 uses 80% less energy than conventional ships and has the ability to travel far and fast on pure electricity. These electric ferries don't need oil changes, so there's no risk of oil leaking into the water. Replacing diesel ferries with electric ones would reduce local emissions, particulate matter, and nitrogen oxide by 100%.

The P-12 is twice as fast as a diesel ferry and 400% more energy efficient than other speedy boats. Candela's goal is to replace their fleet of 70 diesel ferries with electric ferries such as this one in the near future, with pilot programs starting at the beginning of 2023.

EUROSTAR ADDING MORE DESTINATIONS

Eurostar is adding dozens of extra destinations across Europe to create more environmentally friendly travel options!

A bigger and better train network is soon going to connect with the Eurostar brand to make European travel more efficient! The network is currently divided into separate lines for the United Kingdom, the Channel Tunnel, and Europe, but this merger would see these lines united.

The separate lines collectively run 112 trains a day and cater to 18.5 million passengers a year. By improving the efficiency of these journeys, the interconnectedness of these countries will improve through an increased number of direct trains over longer distances.

Dubbed Project Green Speed, these new and improved train lines will encourage travel by train rather than short-haul flights or car rides in the future, which will be better for the environment and faster for passengers. Train travel remains the most environmentally friendly option for travel across Europe, with the lowest greenhouse gas emissions.

The project was initially created in 2019 but it was put on hold amidst the pandemic. Luckily, it is now back and better than ever. Because the united railway service will be run by one brand, there will also be more opportunities for loyalty programs, streamlined transfers, and easier ticketing.

MORE NIGHT TRAINS COMING TO EUROPE

A new fleet of night trains is coming to Europe, so you'll be able to travel more without having to step onto a plane!

They're called *NightJets*, created by Austrian railway company ÖBB, and they're expected to travel to cities including Budapest, Berlin, Milan, Vienna, and many more. Rides on these night trains will be competitively priced to tempt people away from using planes. Sleeping car tickets will be around $53 to $107 (€50 to €100).

The company has been praised for its trailblazing plans to revolutionize the night train scene. Having listened to feedback from frequent travelers and environmental enthusiasts, the trains will focus on comfort, safety, and minimizing its carbon footprint. The company is investing $746 million (€700 million) into 33 NightJet trains that will be launched gradually from the summer of 2023 to 2025. Each train will consist of sleeping carriages, with a variety of sleeping arrangements, and seated carriages.

According to Austria's environment minister, a journey with a NightJet is 50 times more climate-friendly than making the same journey by plane! There are concerns for the success of the night railway, mainly because it faces higher taxes than airlines and railways and has to prioritize shipping goods at night due to the energy crisis. However, the company is confident that the night train niche is one that will only grow from here—especially since it will be beneficial to travelers and most importantly, kinder to our environment.

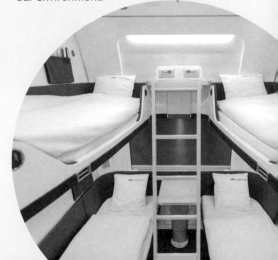

MOMS PUSHING FOR ELECTRIC BUSES

Moms in the United States are fighting for more electric school buses for their children!

The grassroots campaign Clean Buses for Healthy Niños is putting pressure on political leaders to address air pollution. Launched in 2017 by Chispa, a community organizing program of the League of Conservation Voters, the campaign strives to influence governors, school boards, and local leaders to work toward a cleaner environment.

Their activism has helped push the government to announce $17 million in funding to convert diesel school buses to electric and low-emission buses! In addition to the health benefits, each electric bus would save school districts $11,000, too, meaning more money would be available to be put into public schools!

Twenty-five million children ride dirty diesel school buses each day in the United States, exposing them to dangerous toxins that can cause asthma, heart and lung diseases, and cancer. Worse still, students and others in Latinx and low-income communities are at the highest risk of breathing dirty air as they disproportionately suffer from the effects of diesel pollution.

Grassroots campaigns like Clean Buses for Healthy Niños are more important now than ever to ensure children are given a healthy environment to learn and grow, without having to worry about respiratory illnesses.

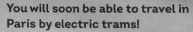

PARIS' ELECTRIC AERIAL TRAMWAY

You will soon be able to travel in Paris by electric trams!

As part of Paris' ecological transformation, a $145 million aerial tramway system is being built, scheduled to open in 2025. The Cable 1 project will connect various suburbs over a 2.8mi (4.5 km) tramway, with travel times predicted to take half the time a bus trip would.

Each cabin will have room for up to 10 passengers. Around 11,000 people per day will be able to make use of the tramway at any of its five stops. To address privacy concerns, the windows of the cable cars will also mist when carriages pass by residential areas!

STEPS YOU CAN TAKE

visit sambentley.co.uk/gnpe/transportation to access relevant links

○ Use public transportation instead of driving. If your community doesn't have accessible and connected public transportation, demand it from your government. Many cities, like Los Angeles in the United States are embracing universal basic mobility as a human right. Take trains and buses for business or vacation travel for a more cost-efficient and environmentally friendly option than air travel and car rentals.

○ Help fight for cleaner school buses and ensure kids get rides to school that don't trigger health problems by supporting Clean Buses for Healthy Niños. Find out how you can call on your local government to make change, too, by visiting the Alliance for Electric School Buses' website.

○ Download some apps to help you travel more efficiently by using public transportation! Citymapper is an app that helps you navigate public transportation, whether you're using a train, subway, bus, ride-sharing service, cab, or bike. Transit is another great app that can automatically display the nearest public transportation options to you. Moovit and Google Maps are other helpful apps to use, too!

CHAPTER 25

Walking & Cycling

Walking, cycling, and skateboarding are low-carbon modes of getting around that don't pollute our environment and, instead, promote social participation in the local economy and community. They offer a wealth of health benefits, too! These nonmotorized transportation modes are inexpensive, flexible, and personal.

Cities investing in walking and cycling infrastructure and policies can not only move closer toward zero-emission transportation, but also can improve their community's air quality, human health, accessibility, and road safety.

Public spaces should be more accessible for everyone. The flow of pedestrian traffic should be eased and cycling lanes should always be an available option. Creating an accessible environment means a wider range of community members can benefit from alternative mobility options. Options like bike sharing can help highly dense cities, too. By using bike sharing, Shanghai saved 8,358 tons of fuel and reduced CO_2 emissions by 25,240 tons and nitrous oxide emissions by 64 tons in 2016.

Simply put, when we have the choice, we can all make a positive impact on our environment by deciding to walk and cycle whenever we can, but it's up to governments to ensure the right infrastructure is in place to make it as accessible as possible. What's not to love about modes that are easy, healthy, fun, and environmentally friendly?

ELECTRIC PAVEMENT

A British company has created tiles that can generate energy from people's footsteps!

Imagine you could help produce clean, off-grid electricity just by walking. Well, the British company Pavegen imagined that, and after 750 prototypes, created a floor tile that could help do exactly that! The triangular tiles can capture people's kinetic energy as they walk and turn it into electricity. One footstep generates around 7 watts, which is enough to light a bulb for 20 seconds.

The tiles have already been installed in 36 countries, including Nigeria, where the technology has been installed on soccer fields, so that when the players run on the field, it actually helps power the lights.

They've also been installed at the University of Birmingham, where students' footsteps help power USB charging at nearby benches.

It's a super-interactive and fun way to generate clean energy that can be used in a bunch of ways on everything from dance floors to train stations and sports fields.

Pavegen hopes their tiles can help people directly engage with clean energy, increase their understanding of sustainability issues, and help be a part of sustainably built environments that empower and connect people.

GREAT AMERICAN RAIL-TRAIL

The United States is building a bike path that goes across the country!

The Great American Rail-Trail will be a multiuse trail that stretches more than 3,700 miles across 12 states, between Washington, D.C., and Washington state.

In 2016, the Rails-to-Trails Conservancy (RTC) began scouting potential routes through Wyoming and Montana and went on to conduct an 18-month route-assessment study that helped identify a preferred route across the United States, looking at existing trails and gaps in the trail. So far the Great American Rail-Trail's preferred route is more than 53% complete!

"Since the 1980s, RTC has understood the potential of a trail like the Great American Rail-Trail that could connect the nation. That vision has been a guidepost for the organization for 30 years. Now, we have the chance to create from that vision, a national treasure that unites millions of people over thousands of miles of trail," said Ryan Chao, president of RTC. "This trail is a once-in-a-lifetime opportunity to provide—together—an enduring gift to the nation that will bring joy for generations to come."

The route will help provide significant economic and social benefits to the communities it connects, and it's predicted it will help create 2,500 jobs.

As the nation's first cross-country, multiuse trail, the Great American Rail-Trail will connect people of all ages and abilities with America's diverse landscapes and communities and help make nature more accessible with nearly 50 million people living within 50 miles (80 kilometers) of the trail.

The Great American Rail-Trail will take several decades to be fully completed, but there's hope the investment of time and resources necessary to complete this trail will be returned many times over as it takes its place among the country's national treasures.

INVESTING IN WALKING & CYCLING

Money is being invested into walking and cycling plans in England, helping millions benefit from cleaner air and cheaper ways to travel!

The new plans—134 in total— will include new footways, cycle lanes, and pedestrian crossings across 46 local authorities outside London, and the government's new executive agency, Active Travel England, will oversee their implementation.

The projects will create new routes and improve existing ones, making it easier and cheaper for people to choose active and green ways of getting around while better connecting communities.

Active Travel Commissioner Chris Boardman said, "This is all about enabling people to leave their cars at home and enjoy local journeys on foot or by bike. Active Travel England is going to make sure high-quality spaces for cycling, wheeling, and walking are delivered across all parts of England, creating better streets, a happier school run, and healthier, more pleasant journeys to work and the shops."

The $248 million (£200 million) budget is part of a wider package of $2.5 billion (£2 billion) that was announced in 2020 to help create a new era for cycling and walking in the United Kingdom.

✓ STEPS YOU CAN TAKE

visit sambentley.co.uk/gnpe/transportation to access relevant links

O Increase active transportation (walking, cycling, or skating) when possible. The World Health Organization (WHO) suggests at least 30 minutes of physical activity per day, and getting around by active transportation is a great way to exercise!

O Contact your law makers and politicians at every level to invest in walking infrastructure and public infrastructure, all of which remains grossly underfunded while car-centered planning and design continue to take the lead.

O Learn more about great urban planning and cities leading the way to help make walking and cycling more accessible, by subscribing to Not Just Bikes on YouTube (@NotJustBikes) and checking out its videos! I recommended watching "Why City Design Is Important (and Why I Hate Houston)."

O Be a part of the movement to complete the Great American Rail-Trail and help speed up the progress of a bike path that spans across the United States! Visit their website to find out more.

OTHER STEPS YOU CAN TAKE

visit sambentley.co.uk/gnpe/othersteps to access relevant links

LAND

○ Sign petitions that push your government to pass policies that regulate farming practices. Petitions that help put mandates in place to minimize deforestation or encourage the planting of trees are especially useful. These mandates will work to help protect the soil we depend on by preventing overfarming and ensuring we are more responsible with our agricultural water usage.

○ The UN observes the World Day to Combat Desertification and Drought every year on June 17. This day is a great opportunity to raise awareness about these important issues. On June 17, post about these issues on social media, tell your friends, and learn more about what is being done. There is strength in numbers, so getting as many people as possible to understand the detrimental effects of deforestation and desertification puts power behind the movement to stop it.

○ As desertification puts greater numbers of people at risk of becoming environmental refugees, you can get involved with groups like the Climate Refugees organization, which work to educate the public and organize think tanks. It provides letter templates you can use to easily contact your government about these issues.

FOOD

○ If you want to go vegan, go at your own pace. If you're trying to move from an omnivorous diet to a vegan diet overnight, it may feel overwhelming. This could mean you're more likely to give up and slip back into old eating habits. You can go at your own pace, making one or two swaps at a time, until you're ready to cut out animal products altogether. Learning how to veganize your favorite meals is a good way to help with the transition!

○ Surround yourself with like-minded people. With social media, finding people who think similarly to you is easy. Change up your feed by following plant-based accounts that inspire you, be it vegan chefs, animal rights activists, or sustainability advocates (or a mix of all three)! This helps create a sense of community and support, especially if you're early on in your vegan journey.

ENERGY

○ Turn down your central heating. The lower the temperature, the more energy (and money) you'll save. Most families are happy with a setting somewhere between 64°F and 70°F (18°C and 21°C), but you may be happy with it being colder or you may need it warmer, especially. if there are any elderly or infirm people living on the property. Remember that different people have different tolerances for cold, so be kind.

○ Install timer plugs. Many devices still use electricity when plugged in, even if they're not being used. Such devices include your TV, microwave, and washing machine. Installing timer plugs, which can be bought online for as little as $6.80 (£5.50), means that you can make sure devices are turned off at times you won't use them without having to think about it.

○ Insulate your house. A huge amount of energy spent heating our homes is wasted due to poor insulation—in a typical British home for example, $1.25 (£1) is wasted for every $3.75 (£3) spent. Improving the insulation of your house prevents heat being lost, giving your central heating system less work to do. Even simple fixes can help. For example, fitting your hot water tank with an insulating jacket can save over $87 (£70) per year. You might like to think about draft-proofing your doors and windows and insulating your floors, walls, roof, tanks, pipes, and radiators. While there are quick wins you can do yourself, many long-term fixes require professional installation.

○ Switch to a renewable-energy provider. Buying electricity from renewable sources will support the growing renewable energy economy. Renewable energy is getting cheaper and—depending on where you live and the energy market at the time—may already be cheaper than energy from fossil fuels, so the switch could save you money. Some energy providers may not provide renewable energy to all customers, but may provide a package that does if you ask for it. Many countries have comparison websites that will help you find a renewable energy provider that's right for you.

CONCLUSION

I don't think I've written a conclusion to something since I was back in secondary school (or "high school" for the American readers among us), but here goes.

Writing this book was a journey. If you're reading this because you've gotten to the end (and you haven't just skipped to the last page), firstly, I want to say thank you so, so much. I feel privileged that you decided to dedicate your time to this book!

Secondly, for me, the point of this book was to introduce people to a diverse variety of topics around sustainability and hopefully light a spark in people's hearts to help them discover and connect with something they really care about. I think most people want to do good and want to play a part in creating a better future but often don't know where to start. If that is you, I hope I've managed to help you find your way!

I'm sure many of the causes and organizations featured in this book would love to hear from you, and there may be ways you can get involved with them, so find them on social media or on their websites, and get involved!

At the time of writing this conclusion, the news has been heavily focusing on the fact that we've just reached 8 billion people on the planet. The tone being used is one that could easily spark division and fear. The way I see it is that we've never had as many people on our planet who can contribute to a brighter future, help regenerate our Earth, come up with solutions for the climate crisis, and be a part of the fight for a better planet.

The future of our planet and the future of all the great things on it doesn't depend on one person flying in and saving the day. It depends on millions, if not billions of us, from every corner of the globe, coming together, showing compassion, and prioritizing unity and love over ego and selfishness. We have the solutions; we just need to implement them.

If you're reading this then there's a very high likelihood you are one of those very important individuals our planet needs right now.

I truly hope this book has helped. Whether that's helping you find a cause to connect with, supporting your mental health, or helping add some balance to the doom and gloom that's thrown at us every day. Remember this book is always here to refer back to should you need a reminder of the good things happening across the world.

There were so many stories we couldn't fit into these pages, so if I've left anything out you really wanted to learn about, I do apologize and promise I did my best to pack in as much as possible! I'm aware some stories will have moved on by the time this book is in your hands, too, but if anything, that just excites me even more to know the good news and innovations covered in this book may be even better by the time you're reading this!

I would love to hear your feedback and your favorite story or cause you came across in this book, so please do share it with me on my social media channels. You can find me @sambentley on most platforms, and be sure to tap that follow button if you haven't already for more good news to pop up in your life!

If you think more people would love to hear good news and would find inspiration or comfort in this book, please leave a positive review on Goodreads or wherever you purchased this book. It'll be greatly appreciated!

Sending best wishes,
Sam

SOURCES

visit sambentley.co.uk/gnpe/sources to access relevant links

OCEANS

16: UNESCO: https://oceanliteracy.unesco.org/ocean-resources/

16: NOAA: https://www.noaa.gov/ocean-coasts

16: Environmental Justice Foundation: https://ejfoundation.org/news-media/ocean-action-is-not-optional-if-we-are-serious-about-ending-the-climate-crisis

16: UNESCO: https://oceanliteracy.unesco.org/ocean-resources/

16: UN Environment Programme: https://www.unep.org/news-and-stories/story/protecting-whales-protect-planet

16: Harvard University: https://sitn.hms.harvard.edu/flash/2019/how-kelp-naturally-combats-global-climate-change/

16: Sabine, C.L., Freely R.A. "The oceanic sink for carbon dioxide" https://www.pmel.noaa.gov/pubs/outstand/sabi2854/sabi2854.shtml

Chapter 1

19: KITV: https://www.kitv.com/news/business/shark-fishing-banned-in-hawaii-in-2022/article_4b07dca6-69b9-11ec-b90b-af8ac98fbb7c.html

19: Oceana: https://europe.oceana.org/importance-sharks-0/

19: Seaspiracy: https://www.seaspiracy.org/

19: Gov.UK: https://www.gov.uk/government/news/government-to-introduce-world-leading-ban-on-shark-fin-trade

20: BBC: https://www.bbc.com/news/world-us-canada-58032702

20: Sky News: https://www.skynews.com.au/australia-news/humpback-whales-no-longer-considered-endangered/video/249ec1b778965904559a039a05b63a88

21: World Wildlife Fund: https://www.wwf.org.au/news/news/2022/wwf-creates-100-000-km2-safe-haven-for-dugongs-in-northern-great-barrier-reef

22: The Guardian: https://www.wwf.org.au/news/news/2022/wwf-creates-100-000-km2-safe-haven-for-dugongs-in-northern-great-barrier-reef

22: World Wildlife Fund: https://www.worldwildlife.org/pages/bait-to-plate

22: CBC: https://www.cbc.ca/news/world/galapagos-illegal-fishing-satellite-technology-1.64694247__vfz=medium%3Dsharebar

22: Open Ocean Robotics: https://openoceanrobotics.com/

23: Billion Oyster Project: https://www.billionoysterproject.org/

Chapter 2

25: The Guardian: https://www.theguardian.com/environment/2021/mar/17/trawling-for-fish-releases-as-much-carbon-as-air-travel-report-finds-climate-crisis

25: Food and Agriculture Organization of the

United Nations: https://www.fao.org/state-of-forests/en/

25: Seaspiracy: https://www.seaspiracy.org/

26: The Guardian: https://www.theguardian.com/environment/2022/feb/07/greenpeace-rocks-sea-bottom-trawling-marine-management-organisation

26: Gov.UK: https://www.gov.uk/government/news/government-uses-brexit-freedoms-to-protect-our-seas

28: Sea Shepherd: https://www.seashepherdglobal.org/who-we-are/

29: Ocean Defenders Alliance: https://www.oceandefenders.org/what-we-do/education-and-outreach.html

Chapter 3

31: World Economic Forum: https://www3.weforum.org/docs/WEF_The_New_Plastics_Economy.pdf

31: Zero Waste Scotland: https://wasteless.zerowastescotland.org.uk/articles/marine-litter

31: Nature: https://www.nature.com/articles/s41598-018-22939-w.pdf

32: Seabin: https://seabin.io/home/

33: The Ocean Cleanup: https://theoceancleanup.com/updates/first-100000-kg-removed-from-the-great-pacific-garbage-patch/

34: Ocean Sole: https://oceansole.com/

35: UN Environment Programme: https://www.unep.org/explore-topics/oceans-seas/what-we-do/addressing-land-based-pollution

LAND

38: World Economic Forum: https://www.weforum.org/agenda/2021/01/earth-surface-ocean-visualization-science-countries-russia-canada-china/

38: Nature Scot: https://www.nature.scot/doc/peatland-action-case-study-whats-connection-between-peat-and-carbon-storage

38: UN Environment Programme World Conservation Monitoring Centre: https://www.unep-wcmc.org/en/news/predicting-the-impact-of-land-use-change-on-biodiversity

Chapter 4

40: World Wildlife Fund: https://support.wwf.org.uk/donate-amazon?pc=DMK004098

40: National Geographic: https://education.nationalgeographic.org/resource/rain-forest

40: World Wildlife Fund: https://www.worldwildlife.org/species/bornean-orangutan

40: World Wildlife Fund: https://www.worldwildlife.org/pages/which-everyday-products-contain-palm-oil#:~:text=Palm%20oil%20is%20the%20most,most%20important%20and%20sensitive%20habitats.

40: Columbus State: http://csc.columbusstate.edu/summers/Outreach/RainSticks/fRainforestFacts.htm

41: The United Nations Office for the Coordination of Humanitarian Affairs: https://reliefweb.int/report/world/world-atlas-desertification-rethinking-land-degradation-

and-sustainable-management

41: Population Resource Bureau: https://www.prb.org/resources/whats-behind-desertification/

41: Zurich Insurance Group: https://www.prb.org/resources/whats-behind-desertification/

41: Food Navigator: https://www.foodnavigator.com/Article/2021/04/26/Regenerative-farming-practices-can-help-restore-the-earth-PepsiCo-discusses-its-Positive-Agriculture-Programme

42: USDA: https://www.usda.gov/media/press-releases/2022/07/25/biden-harris-administration-announces-plans-reforestation-climate

42: AP: https://apnews.com/article/virus-outbreak-india-forests-climate-change-science-fld41fd4772742279da89e972dd8493d

42: BBC: https://www.bbc.com/news/world-africa-49266983

42: Guinness World Records: https://www.guinnessworldrecords.com/world-records/77371-most-trees-planted-in-a-day-by-an-individual

43: Seedballs Kenya: https://www.seedballskenya.com/

44: AirSeed Technologies: https://airseedtech.com/

45: United Nations Convention to Combat Desertification: https://www.unccd.int/our-work/ggwi

Chapter 5

47: Regeneration International: https://regenerationinternational.org/why-regenerative-agriculture/

47: LaCanne, C.E., and Lundgren, J.G. "Regenerative agriculture: merging farming and natural resource conservation profitably": https://www.ncbi.nlm.nih.gov/pmc/articles/PMC5831153/

48: Justdiggit: https://justdiggit.org/

50: World Wildlife Fund: https://www.worldwildlife.org/press-releases/colombia-wwf-and-partners-announce-245m-agreement-to-permanently-protect-vital-systems-of-nation-s-protected-areas

51: Rainforest Trust: https://www.rainforesttrust.org/our-impact/rainforest-news/rainforest-trust-and-its-partners-have-protected-more-than-one-million-acres-to-date-in-2022/

52: Virgin Money: https://uk.virginmoney.com/brighter-money/helping-farmers-create-a-greener-future/

52: Good News Network: https://www.goodnewsnetwork.org/giving-bits-of-farm-land-back-to-nature-does-not-reduce-crop-yields-landmark-study-shows/

53: Justdiggit: https://buy-a-bund.justdiggit.org/

53: Savory: https://savory.global/regenerative-buying-guide/

Chapter 6

54: All Colour Envelopes: https://www.allcolourenvelopes.co.uk/pages/recycling-logos-explained

55: CNN: https://edition.cnn.com/2022/07/01/india/india-bans-single-use-plastic-intl-hnk/index.html

55: Office of Governor Gavin Newsom: https://www.gov.ca.gov/2022/06/30/governor-newsom-signs-legislation-cutting-harmful-plastic-pollution-to-protect-communities-oceans-and-animals/

56: Hoola One: https://hoolaone.com/home/

57: LaserFood: https://laserfood.es/

58: The Guardian: https://www.theguardian.com/environment/2020/oct/30/us-and-uk-citizens-are-worlds-biggest-sources-of-plastic-waste-study

59: UT News: https://news.utexas.edu/2022/04/27/plastic-eating-enzyme-could-eliminate-billions-of-tons-of-landfill-waste/

Chapter 7

61: International Fund for Agricultural Development: https://www.google.com/search?q=ifad&rlz=1C5CHFAenUS918US918&oq=ifad&aqs=chrome..69i57j0i512i5j69i60l2.1398j0j4&sourceid=chrome&ie=UTF-8

62: NPR: https://www.npr.org/2022/05/25/1100823011/native-tribes-celebrate-montana-land-ownership-and-bison-range-restoration

62: National Bison Range: https://bisonrange.org/

62: Grist: https://grist.org/indigenous/rappahannock-tribe-gets-465-acres-land-back-chesapeake-bay/?fbclid=IwAR0SvHMVtFChUG7iG7AjljvFutgviUadGvRZmV1PwH00Z3zZwdeDYmifaEA

63: Queensland Government: https://statements.qld.gov.au/statements/94555#:~:text=The%20Cape%20York%20Peninsula%20Tenure%20Resolution%20Program%20returns%20ownership%20and,and%20cultural%20values%20are%20protected.

64: Washington Post: https://www.washingtonpost.com/climate-environment/2022/06/20/bears-ears-national-monument-tribes/

64: Intergovernmental Cooperative Agreement: https://www.blm.gov/sites/blm.gov/files/docs/2022-06/BearsEarsNationalMonumentInter-GovernmentalAgreement2022.pdf

64: Washington Post: https://www.washingtonpost.com/climate-environment/2022/06/20/bears-ears-national-monument-tribes/

FOOD

68: United Nations: https://www.un.org/development/desa/en/news/population/world-population-prospects-2019.html

68: Forbes: https://www.forbes.com/sites/davidrvetter/2021/03/10/how-much-does-our-food-contribute-to-global-warming-new-research-reveals-all/?sh=4ca7c7db27d7

68: The Guardian: https://www.theguardian.com/environment/2018/nov/28/global-food-system-is-broken-say-worlds-science-academies

68: Chatham House: https://www.chathamhouse.org/2021/02/food-system-impacts-biodiversity-loss

68: The Guardian: https://www.theguardian.com/environment/2018/nov/28/global-food-system-is-broken-say-worlds-science-academies

68: Food and Agriculture Organization of the United Nations: https://www.fao.org/news/story/en/item/197623/icode/

69: University of Oxford: https://www.ox.ac.uk/news/2016-03-22-veggie-based-diets-could-save-8-million-lives-2050-and-cut-global-warming

Chapter 8

70: National Geographic: https://www.nationalgeographic.com/environment/article/141013-food-waste-national-security-environment-science-ngfood#:~:text=The%20Food%20and%20Agriculture%20Organization,processing%20plants%2C%20marketplaces%2C%20retailers%2C

70: Columbia Climate School: https://news.climate.columbia.edu/2019/08/23/food-waste-hidden-costs/

70: World Wildlife Fund: https://www.worldwildlife.org/stories/fight-climate-change-by-preventing-food-waste

70: BBC: https://www.bbc.com/future/article/20200224-how-cutting-your-food-waste-can-help-the-climate#:~:text=One%20third%20of%20greenhouse%20emissions,8%25%20of%20our%20total%20emissions.

71: Hubbub: https://www.hubbub.org.uk/the-community-fridge

72: Edie: https://www.edie.net/wrap-food-redistribution-by-uk-organisations-reached-record-high-in-2021-likely-to-keep-rising/

73: FoodPrint: https://foodprint.org/issues/the-problem-of-food-waste/

74: Food Forward: https://foodforward.org/

Chapter 9

76: Poore, J. "Reducing food's environmental impacts through producers and consumers": https://josephpoore.com/Science%20360%206392%20987%20-%20Accepted%20Manuscript.pdf

77: Oxford University: https://www.ox.ac.uk/news/2016-03-22-veggie-based-diets-could-save-8-million-lives-2050-and-cut-global-warming

78: BBC: https://www.bbc.com/news/business-61565233

79: Plant Based News: https://plantbasednews.org/culture/ethics/cattle-rancher-goes-vegan-shifts-to-plant-based-farming/

80: Independent: https://www.independent.co.uk/climate-change/news/haywards-heath-vegan-plant-based-treaty-b2143892.html

81: Science Daily: https://www.sciencedaily.com/releases/2022/09/220906161505.htm

82: Nature: https://www.nature.com/articles/s41586-022-04629-w

Chapter 10

85: World Food Programme: https://www.wfp.org/stories/5-facts-about-food-waste-and-hunger

85: EPA: https://www.epa.gov/gmi/importance-methane#:~:text=Methane%20is%20more%20than%2025,due%20to%20

human%2Drelated%20activities

88: NPR: https://www.npr.org/2022/02/07/1078777252/a-new-law-in-california-requires-food-waste-to-be-composted

90: The Guardian: https://www.theguardian.com/environment/2022/aug/27/englands-gardeners-to-be-banned-from-using-peat-based-compost

91: New York Times: https://www.nytimes.com/2022/05/13/climate/domingo-morales-composting-nyc.html?smid=fb-nytimes&smtyp=cur

Chapter 11

93: National Geographic: https://education.nationalgeographic.org/resource/earths-fresh-water

93: UN: https://www.un.org/development/desa/en/news/population/world-population-prospects-2019.html

93: National Geographic: https://www.nationalgeographic.com/environment/article/freshwater-crisis

94: MIT News: https://news.mit.edu/2022/portable-desalination-drinking-water-0428

95: CloudFisher: https://www.wasserstiftung.de/en/cloudfisher-extracting-clean-water-from-fog-with-innovative-technology/

95: UN Women: https://www.unwomen.org/en/news/stories/2018/8/speed-ded-regner-stockholm-world-water-week

96: Warka Water: https://warkawater.org/

96: Water.org: https://water.org/our-impact/water-crisis/#:~:text=Today%2C%20771%20million%20people%20%20E2%80%93%20301,are%20the%20people%20we%20empower.

98: UT News: https://news.utexas.edu/2022/05/23/low-cost-gel-film-can-pluck-drinking-water-from-desert-air/

99: The Wildlife Trusts: https://www.wildlifetrusts.org/actions/how-conserve-water

99: Waterwise: https://www.waterwise.org.uk/save-water/

WILDLIFE

102: World Wildlife Fund: https://www.wwf.org.uk/what-we-do/protecting-wildlife

102: National Geographic: https://education.nationalgeographic.org/resource/wildlife-conservation

102: BBC: https://www.bbc.com/news/science-environment-54091048

102: Stiles, William "The importance of biodiversity and wildlife on farmland": https://businesswales.gov.wales/farmingconnect/sites/farmingconnect/files/documents_and_files/Lluniau_2017/technical_article_-_farmland_biodiversity_final.pdf

102: University of Oxford: https://www.ox.ac.uk/news/2022-03-03-large-mammals-can-help-climate-change-mitigation-and-adaptation

102: BBC Earth: https://www.bbcearth.com/news/what-can-animals-teach-us-about-mental-health

Chapter 12

104: The Guardian: https://www.theguardian.com/environment/2020/feb/17/

beavers-cut-flooding-and-pollution-and-boost-wildlife-populations

104: National Geographic: https://www.nationalgeographic.com/animals/article/yellowstone-wolves-reintroduction-helped-stabilize-ecosystem

105: Ark Nature; https://www.ark.eu/en/projects/european-bison

106: Rewilding Institute: https://rewilding.org/once-extinct-macaws-are-repopulating-ibera/

106: Rewilding Argentina: https://www.rewildingargentina.org/

Chapter 13

108: National Geographic: https://education.nationalgeographic.org/resource/conservation

108: Conservation Handbook: https://www.conservationhandbooks.com/why-conserve/

108: World Wildlife Fund: https://www.worldwildlife.org/threats

109: Highland Wildlife Park: https://highlandwildlifepark.org.uk/news/article/20980/highland-wildlife-park%E2%80%99s-adorable-snow-leopard-cubs-named/

110: World Wildlife Fund: https://app.yearly.report/newbuilder/#/from/wwf/wwfs-impact-on-tiger-recovery

112: The Guardian: https://www.theguardian.com/environment/2022/jun/05/whale-watching-season-starts-early-as-humpback-population-bounces-back

113: FOX 2 Detroit: https://www.fox2detroit.com/news/river-otter-spotted-in-detroit-river-may-be-1st-sighting-in-100-years.amp

114: NBC: https://www.nbcnews.com/news/asia/rare-birth-endangered-sumatran-rhino-sparks-hope-conservation-efforts-rcna22336

114: World Wildlife Fund: https://www.worldwildlife.org/species/sumatran-rhino

115: BBC: https://www.bbc.co.uk/newsround/60602335

Chapter 14

117: Britannica: https://www.britannica.com/explore/savingearth/the-hidden-history-of-greco-roman-vegetarianism

117: World Animal: https://worldanimal.net/our-programs/strategic-advocacy-course-new/module-1/history

117: The Guardian: https://www.theguardian.com/environment/2018/may/31/avoiding-meat-and-dairy-is-single-biggest-way-to-reduce-your-impact-on-earth?CMP=fb_gu

118: Gov.UK: https://www.gov.uk/government/news/lobsters-octopus-and-crabs-recognised-as-sentient-beings

120: BBC: https://www.bbc.com/news/science-environment-62213459?at_medium=RSS

120: Gov.UK: https://www.gov.uk/government/news/eurasian-beavers

120: Natural History Museum: https://www.nhm.ac.uk/discover/news/2022/march/beavers-reintroduced-to-london-after-400-years.html

121: Inside Climate News: https://insideclimatenews.org/news/29032022/

ecuadors-high-court-rules-that-wild-animals-have-legal-rights/

122: Forbes: https://www.forbes.com/sites/nicholasreimann/2022/01/21/new-law-banning-dog-chaining-takes-effect-in-texas-and-dominates-on-facebook/?sh=4aa9e1ea1dd7

122: Texas Humane Legislation Network: https://www.thln.org/a_brand_new_day

123: PETA UK: https://www.peta.org.uk/blog/italy-fur-ban/

124: Plant Based News: https://plantbasednews.org/culture/ethics/horse-drawn-carriages-banned-mallorca/

124: PETA: https://secure.peta.org.uk/page/30696/action/1?ea.url.id=6011680&forwarded=true

CITY

128: HuffPost: https://www.huffpost.com/entry/big-cities-benefit-environment_n_5d1b99c2e4b07f6ca5851593

128: Bloomberg: https://www.bloomberg.com/news/articles/2012-04-19/why-bigger-cities-are-greener#:~:text=Larger%2C%20denser%20cities%20are%20cleaner,human%20encroachment%20on%20natural%20habitats.

128: World Economic Forum: https://www.weforum.org/agenda/2022/04/cities-driving-climate-change-but-part-of-the-solution-un-report/

128: Bloomberg: https://www.bloomberg.com/news/articles/2012-04-19/why-bigger-cities-are-greener#:~:text=Larger%2C%20denser%20cities%20are%20cleaner,human%20encroachment%20on%20natural%20habitats.

128: United Nations: https://www.un.org/development/desa/pd/file/1942#:~:text=By%202050%2C%2068%25%20of%20world,World%20Urbanization%20Prospects%20%7C%20Population%20Division

Chapter 15

130: IW: https://iwaponline.com/bgs/article/2/1/1/71141/Urban-agriculture-as-a-keystone-contribution

130: USDA: https://www.usda.gov/media/press-releases/2022/02/01/usda-announces-inaugural-federal-advisory-committee-urban

130: Climate ADAPT: https://climate-adapt.eea.europa.eu/en/metadata/adaptation-options/urban-farming-and-gardening#:~:text=Areas%20used%20for%20urban%20farming,for%20allotment%20and%20vegetable%20gardens.

130: National Library of Medicine: https://www.ncbi.nlm.nih.gov/pmc/articles/PMC7558991/

131: TimeOut: https://www.timeout.com/news/the-worlds-largest-vertical-farm-is-being-built-in-the-uk-060922

132: VegNews: https://vegnews.com/2022/7/Eben-Bayer-MyForest-Foods-vegan-bacon

133: EuroNews: https://www.euronews.com/green/2022/05/25/brussels-grows-fruits-and-vegetables-on-top-of-a-supermarket

134: Grist: https://grist.org/agriculture/in-detroit-a-push-to-help-black-farmers-purchase-land/

135: Bloomberg: https://www.bloomberg.com/

news/articles/2022-08-21/rio-de-janeiro-brazil-set-to-build-world-s-largest-urban-garden

136: BBC: https://www.bbc.com/worklife/article/20201214-how-15-minute-cities-will-change-the-way-we-socialise

Chapter 16

137: Food and Agriculture Organization of the United Nations: https://www.fao.org/news/story/en/item/1194910/icode/#:~:text=Bees%20and%20other%20pollinators%20such,plus%20many%20plant%2Dderived%20medicines.

137: Our World in Data: https://ourworldindata.org/habitat-loss

137: World Wildlife Fund: https://wwf.panda.org/discover/our_focus/wildlife_practice/problems/habitat_loss_degradation/

137: University of Sussex: http://www.sussex.ac.uk/lifesci/goulsonlab/blog/bee-friendly-flowers

138: Dezeen: https://www.dezeen.com/2022/01/24/bee-bricks-planning-requirement-brighton/

139: HuffPost: https://www.huffingtonpost.co.uk/entry/detroit-hives-honeybees_n_5a6cfc9ee4b0ddb658c7019c

140: Reasons to be Cheerful: https://reasonstobecheerful.world/the-buzzing-efforts-to-save-the-bees-in-the-netherlands/

Chapter 17

144: World Health Organization: https://www.who.int/health-topics/floods#tab=tab_1

144: OECD: https://read.oecd-ilibrary.org/finance-and-investment/financial-management-of-flood-risk_9789264257689-en#page14

144: Flood Guidance: https://www.floodguidance.co.uk/about-us/

145: World Health Organization: https://www.who.int/health-topics/floods#tab=tab_1

147: Reasons to be Cheerful: https://reasonstobecheerful.world/pakistans-mangroves-coastal-conservation-climate-change/

Chapter 18

149: National Library of Medicine: https://www.ncbi.nlm.nih.gov/pmc/articles/PMC5663018/

149: CityChangers: https://citychangers.org/sustainable-buildings/greenery/

149: Ciria: http://www.opengreenspace.com/opportunities-and-challenges/climate-change/flood-management/

149: New Scientist: https://www.newscientist.com/article/2298675-trees-cool-the-land-surface-temperature-of-cities-by-up-to-12c/#:~:text=The%20cooling%20effect%20of%20trees,cities%20adapt%20to%20global%20warming.

149: New York University: https://wp.nyu.edu/sustainableinfrastructure/2018/12/07/greenifying-the-skyline/

150: Scotscape: https://www.scotscape.co.uk/services/living-pillar

151: Leicester City Council: https://news.leicester.gov.uk/news-articles/2021/may/new-network-of-living-roof-bee-friendly-bus-stops-springing-up-in-leicester/

152: World Economic Forum: https://www.

weforum.org/videos/madrids-wind-garden-will-cool-city-by-4-degrees

154: The Guardian: https://www.theguardian.com/environment/2022/aug/04/walking-forest-of-1000-trees-transforms-dutch-city-aoe

ENERGY

158: Our World in Data: https://ourworldindata.org/energy-mix

158: Oxfam: https://www.oxfam.org/en/press-releases/carbon-emissions-richest-1-percent-more-double-emissions-poorest-half-humanity

Chapter 19

161: Independent: https://www.independent.co.uk/tech/solar-power-renewable-energy-europe-b2036988.html

162: NPR: https://www.npr.org/2022/04/07/1091320428/solar-panels-that-can-generate-electricity-at-night-have-been-developed-at-stanf

163: Reuters: https://www.reuters.com/business/energy/portugal-set-start-up-europes-largest-floating-solar-park-2022-05-09/

164: Independent: https://www.independent.co.uk/climate-change/news/solar-panels-new-buildings-eu-mandatory-b2081732.html

165: Good News Network: https://www.goodnewsnetwork.org/delaware-gives-free-solar-to-low-income-residents/

166: Good News Network: https://www.goodnewsnetwork.org/california-begins-covering-canals-with-solar-panels-to-fight-drought/

167: Smartflower: https://smartflower.com/

Chapter 20

170: Recharge: https://www.rechargenews.com/transition/tidal-power-impact-on-marine-wildlife-significantly-lower-than-predicted-study/2-1-768209

170: Eco Wave Power: https://www.ecowavepower.com/our-technology/how-it-works/

171: Sea Wave Energy Ltd: https://swel.eu/

172: Energy.gov: https://www.energy.gov/articles/doe-announces-25-million-cutting-edge-wave-energy-research

Chapter 21

174: May, Roel, et al. "Paint it black": https://onlinelibrary.wiley.com/doi/epdf/10.1002/ece3.6592

174: Energy Monitor: https://www.energymonitor.ai/tech/renewables/weekly-data-how-many-birds-are-really-killed-by-wind-turbines#:~:text=Merriman%20concludes%20that%201.17%20million,in%20the%20US%20each%20year.&text=This%20is%20a%20lot%20of,that%20live%20in%20the%20US.

175: Good News Network: https://www.goodnewsnetwork.org/zebra-general-electric-100-percent-recyclable-wind-turbine-blade/

176: Tulip Turbines: https://flowerturbines.com/

177: The Verge: https://www.theverge.

com/2022/2/11/22929059/recycled-wind-turbine-blade-bridges-world-first

178: Euro News: https://www.euronews.com/green/2022/03/17/italy-builds-six-new-wind-farms-in-a-bid-to-move-away-from-russian-gas-dependency

179: BBC: https://www.bbc.com/news/uk-wales-62037754

Chapter 22

181: World Economic Forum: https://www.weforum.org/agenda/2022/07/electric-vehicles-world-enough-lithium-resources/

181: Blakers, Andrew, et al. "A review of pumped hydro energy storage": https://iopscience.iop.org/article/10.1088/2516-1083/abeb5b

182: iGem: https://www.igem.org.uk/news-and-publications/igem-news/hydrogen-is-crucial-to-reaching-net-zero-says-climate-change-committee-chief/#:~:text=Climate%20Change%20Committee%20CEO%20Chris,country%20over%20the%20coming%20years.

183: BBC: https://www.bbc.com/news/science-environment-61996520

184: Renew Economy: https://reneweconomy.com.au/australian-start-up-eyes-disused-mine-shafts-for-giga-scale-gravity-energy-storage/amp/

184: Green Gravity: https://greengravity.com/technology/

TRANSPORTATION

188: EPA: https://www.epa.gov/greenvehicles/greenhouse-gas-emissions-typical-passenger-vehicle#:~:text=typical%20passenger%20vehicle%3F-,A%20typical%20passenger%20vehicle%20emits%20about%204.6%20metric%20tons%20of,8%2C887%20grams%20of%20CO2.

188: ICCT: https://theicct.org/publication/co2-emissions-from-commercial-aviation-2018/

188: Kärcher, Bernd. "Formation and radiative forcing of contrail cirrus." https://www.nature.com/articles/s41467-018-04068-0

Chapter 23

190: ICCT: https://theicct.org/sites/default/files/publications/Global-LCA-passenger-cars-jul2021_0.pdf

191: Fortune: https://fortune.com/longform/blood-sweat-and-batteries/

191: Business & Human Rights Resource Centre: https://www.business-humanrights.org/en/latest-news/tracking-human-rights-violations-environmental-impacts-in-lithium-batteries-supply-chains-in-china-drc-so-america/#c149788

192: Bloomberg: https://www.bloomberg.com/news/articles/2022-06-01/oil-s-displacement-as-a-road-fuel-is-about-to-ramp-up-bnef-says?leadSource=uverify%20wall

193: ABB: https://www.bloomberg.com/news/articles/2021-12-09/peak-oil-demand-is-coming-but-not-so-soon#xj4y7vzkg

Chapter 24

196: ScienceDirect: https://www.sciencedirect.com/science/article/pii/

S266701002100054

8#:~:text=Public%20transportation%20can%20decrease%20the,public%20transportation%20for%20congestion%20reduction.

197: Candela: http://candela.com/p-12-shuttle/

198: TimeOut: https://www.timeout.com/news/eurostar-will-soon-be-linked-up-with-dozens-of-extra-destinations-across-europe-042622

198: UIC: https://uic.org/com/enews/nr/662/article/france-belgium-green-speed-a-project-to-combine-eurostar-and-thalys-has-been

199: The Guardian: https://www.theguardian.com/world/2022/sep/08/europes-next-generation-night-trains-aim-to-draw-passengers-away-from-planes

200: The 19th: https://19thnews.org/2022/03/moms-electric-school-buses-communities-children-health/

200: Chispa: https://chispalcv.org/clean-buses-for-healthy-ninos/

201: CNN: https://www.cnn.com/travel/article/paris-to-build-145m-cable-car-system/index.html#:~:text=Paris%20connection,open%20in%20Paris%20by%202025.&text=Cable%201%20(C1)%20has%20an,between%20cabins%20at%20peak%20times.

202: EuroNews.next: https://www.euronews.com/next/2022/05/20/universal-basic-mobility-cities-tackle-the-transport-gap-with-free-transit-e-bikes-and-car#:~:text=LA%20is%20just%20one%20of,their%20socioeconomic%20status%20or%20disabilities.

Chapter 25

203: UN Environment Programme: https://www.unep.org/resources/report/share-road-global-outlook-walking-and-cycling-october-2016?_ga=2.220665823.613264663.1662297890-684793798.1659082489

203: Zhang, Yongping, et al. "Environmental benefits of bike sharing": https://www.sciencedirect.com/science/article/abs/pii/S0306261918304392

204: Pavegen: https://www.pavegen.com/

205: Rails to Trails: https://www.railstotrails.org/greatamericanrailtrail/

206: Gov.UK: https://www.gov.uk/government/news/healthy-cost-effective-travel-for-millions-as-walking-and-cycling-projects-get-the-green-light

CONCLUSION

208: Climate Refugees: https://www.climate-refugees.org/

209: BBC: https://www.bbc.co.uk/news/technology-61235367.amp

209: RBKC: https://www.rbkc.gov.uk/environment/climate-change/heat-loss-your-home

Sources **215**

PHOTO CREDITS

Ministry of Environment and Forestry; **115 (top),** Svitlana Tkach/Shutterstock; **115 (bottom),** Ian Duffield/Shutterstock; **115 (inset),** Emblan/Shutterstock; **118 (top),** Vladimir Wrangel/Shutterstock; **118 (mid),** Jesus Cobaleda/Shutterstock; **118 (bottom),** RLS Photo/Shutterstock; **119,** Courtesy of Edge Innovations; **120,** Danny Iacob/Shutterstock; **121,** Zaruba Ondrej/Shutterstock; **121 (inset),** Helge Zabka/Shutterstock; **122 (top),** paul prescott/Shutterstock; **122 (bottom),** NReiher/Shutterstock; **123,** Algimantas Barzdzius/Shutterstock; **123 (inset),** Polawat Klinkulabhirun/Shutterstock; **124 (top),** Nataliya Schmidt/Shutterstock; **124 (bottom),** Eyuel Melese Muse/Shutterstock; **125,** a katz/Shutterstock; **129,** Mehmet Recep Ozdemir/Shutterstock; **131,** Courtesy of Jones Food Co.; **132,** Photo by Mitch Wojnarowicz/Courtesy of MyForest Foods; **133,** © European Union/Fred Guerdin; **134,** Joshua Resnick/Shutterstock; **135,** Marco Antonio Rezende/Prefeitura do Rio; **136,** James Andrews1/Shutterstock; **138,** Courtesy of Green&Blue; **139,** Courtesy of Detroit Hives; **140-141,** Mike Wiering/Shutterstock; **142,** Chatham172/Shutterstock; **142 (inset),** Sabine Seiter_sh/Shutterstock; **146,** Courtesy of Høje Taastrup C; **150,** Courtesy of Scotscape; **151,** Courtesy of Leicester City Council; **152-153,** © West 8; **154,** Courtesy of Arcadia; **155,** R. de Bruijn_Photography/Shutterstock; **161 (top),** Make more Aerials/Shutterstock; **161 (bottom),** Make more Aerials/Shutterstock; **162 (top),** franconiaphoto/Shutterstock; **162 (bottom),** zhengzaishuru/Shutterstock; **163,** Courtesy of EDP; **164-165,** FotoHelin/Shutterstock; **165,** surasak jailak/Shutterstock; **166,** Courtesy of Solar Aquagrid; **167,** Courtesy of Smartflower; **168,** Alessandro28/Shutterstock; **170,** Courtesy of Eco Wave Power; **171,** Courtesy of SWEL; **172,** Courtesy of CalWave; **173,** Courtesy of Eco Wave Power; **175,** Courtesy of General Electric; **176,** Courtesy of Flower Turbines; **177,** Courtesy of Re-Wind Network; **177 (inset),** ka100500/Shutterstock; **178,** Courtesy of Renexia; **179,** Courtesy of Principle Power; **180,** Vladimka production/Shutterstock; **182,** r.classen/Shutterstock; **183,** Courtesy of Polar Night Energy; **185,** Courtesy of Green Gravity; **192,** buffaloboy/Shutterstock; **193,** Courtesy of ABB; **194,** Obs70/Shutterstock; **197,** Courtesy of Candela; **198,** olrat/Shutterstock; **199,** Courtesy of Siemens; **200,** Courtesy of Zūm; **201,** © Île-de-France Mobilités; **204,** Courtesy of Pavegen; **205,** Courtesy of Great American Rail Trail; **206,** ironbell/Shutterstock

INDEX

15-minute city (Paris), 136

A

active transport, 207
Active Travel England, 206
Agri E Fund (Virgin Money), 52
agroecology, 47
agroforestry, 39
AirSeed Technologies, 44
Amazon rainforest,
  deforestation, 40
Animal Equality Italy!, 123
Animal Liberation (Singer), 117
Animal Rebellion, 125
animal rights, 117
  beavers, protecting, 120
  chained dogs in Texas, 122
  harmful practice bans, 123
  horse-drawn carriages, 124
  robot dolphins, 119
  sentient being classification, 118
Animal Welfare Bill, 118
Avant-Garde Vegan, 84

B

batteries, sand batteries, 183
Bayer, Eben, 132
beaches, HO Micro beach
  vacuum, 56
Bears Ears National Monument, 64
Beaver Trust, 107
beavers, 104, 120
Bee Bus Stops, 151
bee habitats, 137
  bee bricks, 138
  Bee Bus Stops, 151
  Build for Bees bee houses, 142
  Detroit hives, 139
  mowing and, 143
  Netherlands National
    Pollinator Strategy, 140
  pollinator-friendly planting, 143
Before the Flood, 11, 76, 84
Best Before Dates, food waste
  and, 73

bicycling, 191, 194. *See also*
  cycling infrastructure
Billion Oyster Project, 23, 24
biodiversity, land, 38–39
bison, European,
  reintroduction, 105
BOSH!, 84
bottom trawling the ocean,
  Greenpeace and, 26
Bowekaty, Carleton, 64
Build for Bees bee houses, 142
bulk shopping, 60
bunds, 48
Bureau of Land Management, 64

C

Campaign Against Living
  Miserably, 10
car sharing, 195
carbon emissions, soil, 38
carpooling, 195
cattle, land use and, 77
Cheap Lazy Vegan, 84
city, 128–129
  15-minute city, 136
  bee habitats, 137–142
  flood prevention, 144–147
  green public spaces, 149–154
  urban farms, 130–136
Clean Buses for Healthy Niños, 200
Climate Refugees, 208
climate-native grass, 148
CloudFishers, 95
coastal floods, 144
community fridges, 71, 75
Compost Power, 91
composting, 92
  as burial/cremation
    alternative, 89
  food waste and, 88
  materials, 87
  Morales, Domingo, 91
  peat, banning, 90
  pH balance, 86
  types, 85–86
conservation, 108
  Golden Eagles, 115
  humpback whales, 112
  marine wildlife, 20–21

river otters, 113
sharks, 19
Sumatran rhino, 114
tiger populations, 110–111
wildcats, 109
Copenhagen skate park, 146
Cowspiracy, 76, 84
cycling infrastructure, 203
  Great American Rail-Trail, 205
  investing in, 206

D

dairy and meat
  cattle, land use and, 78
  FFSFF (Farmers For Stock-Free
    Farming), 78
  Before the Flood, 76
  meat alternatives, 82
  Plant Based Treaty, 80
  plant-based meats, 81
  veganism, 78–81
De Gelder, Paul, 19, 24
debris removal from sea, 29
decarbonization, 159
deforestation, 40–46. *See also*
  planting projects
  dairy and meat production, 82
  mangrove rehabilitation, 147
  tree planting projects, 42
  woodland protection, 46
desalination of water, 94
desertification, 41–46. *See also*
  planting projects
  World Day to Combat
    Desertification and Drought, 208
Detroit Black Farmer Land
  Fund, 134, 136
Detroit River otters, 113
DiCaprio, Leonardo, 76
diet, plant-based, 30
dogs, chaining bans, 122
dolphins, robot, 119
Dominion, 76
doorstep garden, 136

E

Earthling Ed, 125
Earth's surface area
  land, 38

ocean, 16
water, 93
Eco Wave Power, 170
Eden Reforestation Project, 46
Edge Innovations, 119
electric bikes, 191
  incentives, 194
electric buses, 200
electric cable cars, 201
electric ferry, 197
electric pavement, 204
electricity. See also renewable
  energy
emissions, food and, 68
energy, 158–159. See also
  renewable energy steps to
  take, 209
energy storage, 181–184
Eshel, Gidon, 76, 84
Eurostar, 198
EVs (electric vehicles), 190–193

F

farmers markets, 60
farming. See also urban farms
FFSFF (Farmers For Stock-Free
  Farming), 78
Fisheries Act, 26
flash floods, 144
flip flop recycling, 34
flood prevention, 144–147
flood types, 144
flower wind turbines, 176, 180
food
  dairy and meat, 76–84
  emissions and, 68
  rerouting from landfills, 74
  steps, 208
Food Forward, 74
food waste, 70–75
forestry
  agroforestry, 39
  deforestation, 40–46
fossil fuels, 158

G

Garden of Wind (el Jardin del
  Viento), 152
Global Ocean Treaty, 30
Golden Eagles, 115

gravitational energy storage,
  184
Great American Rail-Trail, 205,
  207
Great Barrier Reef, 21
Great Green Wall, 45
Great Pacific Garbage Patch,
  31, 33
Green Dot label, plastic
  pollution, 54
Green Gravity, 184
green public spaces, 149
  Bee Bus Stops, 151
  el Jardin del Viento (Garden of
    Wind), 152
  LivingPillar, 150
  walking forest, 154
Green&Blue, 138
Greenery Skyrise Incentive, 149
Greenpeace, bottom trawling
  and, 26
groundwater floods, 144

H

Haywards Heath, veganism, 80
Highland Wildlife Park, 109
HO Micro beach vacuum, 56
Hoola One, 56
Hopi Tribe, 64
horse-drawn carriages, 124
Hortas Cariocas project (Rio
  de Janeiro), 135
Hubbub community fridge, 71
Huffstetler, Emily, 142
humpback whales, 112

I

illegal fishing, 22
indigenous communities
  Bears Ears National
    Monument, 64
  Confederated Shalish, 62
  Hopi Tribe, 64
  Indigenous Knowledge
    Systems, 60
  Kootenai, 62
  Navajo Nation, 64
  Pueblo of Zuni, 64
  Rappahannock, 62
  returning tribal homelands,
    62–63

Uintah and Ouray Reservation,
  64
Ute Tribe, 64
Indigenous Knowledge
  Systems, 60
Interceptor Trashfence, 33
Italy, harmful practice bans,
  animals, 123
IUCN Red List of Threatened
  Species, 105

J

Juniper, Tony, 120
Justdiggit, 53

K

Kent Wildlife Trust, 105
King-Sonnen, Renee, 79

L

land
  agroforestry, 39
  Bears Ears National
    Monument, 64
  biodiversity, 38–39
  composting, 85–92
  deforestation, 40–46
  desertification, 41–46
  Earth's surface area, 38
  meat production, 77
  milk production, 77
  peatland, 38
  plastic pollution, 54–59
  returning tribal homelands,
    62–63
  soil, 38
  steps to take, 208
land protection, 50–51
Land to Market verified seal, 53
LaserFood, 57
laser-fruit markings, 57
Leck, Richard, 21
litter, ocean cleanup and, 31–35
LivingPillars, 150
Lloyd-Jones, James, 131

M

macaws, reintroducing, 106
mangrove forests, 147
meat alternatives, 82

Merrell, Matt, 185
methane, food waste, 70
Mobius Loop label, plastic
  pollution, 54
Modern Scientific Knowledge
  System, 60
Moses, Antoine, 42
MPAs (Marine Protected
  Areas), 26
mushroom farm, vertical, 132
MyForest Foods, 132

N

NACUE (National Association
  of College & University
  Entrepreneurs), 9
natural branding, 57
Navajo Nation, 64
Noad, Mike, 112

O

Ocean Cleanup, 33, 35
Ocean Sole, 34, 35
oceans
  Billion Oyster Project, 23
  bottom trawling, 25–30
  cleanup innovations, 31–35
  CO2 absorption, 16
  debris removal, 29
  desalination, 94
  Earth's surface area, 16
  Global Ocean Treaty, 30
  Great Barrier Reef, 21
  Great Pacific Garbage Patch,
    31, 33
  humpback whales, 20, 112
  illegal fishing, 22
  Interceptor Trashfence, 33
  as life support for planet, 17
  marine wildlife conservation,
    20–21
  Ocean Cleanup, 33
  Ocean Sole, 34
  Open Ocean Robotics, 22
  oxygen production, 16
  plastic pollution, 31–32
  robot dolphins, 119
  Sea Shepherd, 28
  Seabin Project, 32
  Seaspiracy, 30

unregulated fishing, 28
ODA (Ocean Defenders
  Alliance), 29
Open Ocean Robotics, 22
Orangutan populations, 40
Oxford Martin School, 77
oxygen, ocean producing, 16
oyster reefs, 23

P

P-12 electric ferry, 197
palm oil, deforestation and, 46
Pavegen, 204
peatland, 38
  protecting, 90
PETA, 125
Plant Based Treaty, 80
plant-based diets, 30
  meats, 81
planting
  climate-native grass, 148
  Great Green Wall, 45
  LivingPillars, 150
  seedball planting, 43
  seed-firing drones, 44
  tree planting, 42
plastic pollution, 31–32, 54–59
plastic-eating enzyme, 59
poaching
  illegal fishing, 22
  unregulated fishing, 28
pollution. *See also* plastic
  pollution
  Great Pacific Garbage Patch,
    31, 33
  Interceptor Trashfence, 33
  Seabin Project, 32
Poore, Joseph, 76
population growth, world, 128
product labeling, 46
public spaces, greenifying, 149
public transportation, 196
  electric buses, Clean Buses
    for Healthy Niños, 200
  electric cable cars in Paris, 201
  electric ferry, 197
  Eurostar, 198
  night trains in Europe, 199
Pueblo of Zuni, 64

R

Rachel Ama, 84
rainwater
  bunds, 48
  harvesting, 148
  skatepark drainage, 146
Rancher Advocacy Program,
  79
Recompose, 89
recycling, 35
  labels, 54
  Ocean Sole, 34
  plastic-eating enzyme, 59
  secondhand shops, 60
red and green macaws,
  reintroducing, 106
regenerative agriculture, 47, 52
  bunds, 48
  Justdiggit, 53
  land protection, 50–51
  wildfire prevention, 49
reintroducing species, 104–106
renewable energy, 158–159
  photo-voltaic cells, 160
  solar power, 160–168
  tidal energy, 169–172
  wind turbines, 174–179
Resin Identification Codes 1-7,
  54
rhinoceros, Sumatran,
  conservation, 114
river floods, 144
river otters, 113
robot dolphins, 119
Ron Finley project, 155
rooftop gardens, 148
rooftop gardens, 133
Rowdy Girl Ranch, 79

S

sand batteries, 183
Saving Wildcats, 109
Scarborough, Peter, 81
Sea Shepherd, 28
sea wave energy, 171
Sea Wave Energy Ltd, 171
Seabin Foundation, 35
Seabin Project, 32
Seaspiracy, 30, 84

secondhand shops, 60
seed pod planting drones, 44
seedball planting, 43
Seedballs Kenya, 43, 46
seed-firing drones, 44
sentient being classification of animals, 118
sewer floods, 144
shark fin bans, 19
sharks, protecting, 19
Shone, Ella, 58
Singer, Peter, 117
single-use plastic bans, 55
SmartFlower solar panels, 167
soil, 38
solar power, 160–168
South of Scotland Golden Eagle Project, 115
Spade, Katrina, 89
Springmann, Marco, 69
Sumatran Orangutan Conservation Program, 107
Sumatran rhino, 114

T

Tagelagi, Dalton, 22
Terra 360 EV charger, 193
Texas, chained dogs bans, 122
The Dark Truth Behind Tourism, 10
tidal energy
   sea wave energy, 171
     Waveline Magnet, 171
   tidal barrages, 169
   tidal lagoons, 169
   tidal streams, 169
   wave energy converter, 170
   wave energy research, 172
tiger conservation, 110–111
Top-up Truck, 58
transportation, 188
   active transport, 207
   car sharing, 195
   carpooling, 195
   cycling infrastructure, 203–206
   electric bikes, 191, 194
   EVs (electric vehicles), 190–193
   passenger vehicles, 188
   public transportation, 188, 196–200

walking infrastructure, 203–206
ZipCar, 194
trashfence. See Interceptor Trashfence
Traylor, Cindy, 79
Traylor, Richard, 79
tree planting, 42, 46
tulip wind turbines, 176

U

Uintah and Ouray Reservation, 64
UK Centre for Ecology & Hydrology, 52
UNILAD, 9–10
University of Texas, plastic eating enzyme, 59
urban farms, 130
   Detroit Black Farmer Land Fund, 134, 136
   doorstep garden, 136
   Hortas Cariocas project (Rio de Janeiro), 135
   mushroom farm, vertical, 132
   rooftop gardens, Brussels, 133
   vertical, 131
US Forest Service, 64
Ute Tribe, 64

V

Varsity Pitch competition, 10
veganism
   farmers, 78–79
   FFSFF (Farmers For Stock-Free Farming), 78
   Haywards Heath, 80
   Plant Based Treaty, 80
   research, 84
   shifting to, 208
   YouTube chefs, 84
vertical farms, 131
   mushroom farm, 132
Virgin Money Agri E Fund, 52
volunteer opportunities, ocean, 24

W

walking forest, 154
walking infrastructure, 203
   electric pavement, 204

investing and, 206
Warka Water, 96
water
   CloudFishers, 95
   conserving, 100
   desalination, 94
   from dryland, 98
   Earth's surface area, 93
   fog collectors, 95
   global crisis, 93
   Warka Water, 96
wave energy converter, 170
Wave Energy Converter (WEC), 171
Waveline Magnet, 171
West 8 architecture firm, 152
What the Health, 84
wildcat conservation, 109
wildfires, prevention, goats and, 49
wildlife, 102–103
   animal rights, 117–124
   conservation, 108–115
   IUCN Red List of Threatened Species, 105
   ocean as home, 16
   reintroducing species, 104–105
Wildwood Trust, 105
wind turbines
   birds and, 174
   horizontal axis, 174
   offshore wind farms, 178
   recyclable blades, 175
   tulip turbines, 176
   upcycling, 177
   vertical axis, 174
   Wales floating wind farm, 179
   Zero wastE Blade ReseArch (ZEBRA), 175
woodland protection, 46
World Animal Protection, 10
World Day to Combat Desertification and Drought, 208
world population growth, 128
WRAP, food redistribution, 72

X–Y–Z

ZipCar, 194

SAM BENTLEY is a climate activist and sustainability content creator who's worked in social media for over a decade. He has built some of the largest platforms across social media, reaching millions of people. His main goal is to use his social media channels to inspire and create positive change. You can follow him on Instagram and TikTok @sambentley. Sam now creates accessible content around climate solutions and spreads awareness about meaningful initiatives happening around the world to help secure a safe and inclusive future for the following generations. He lives in Birmingham, England.

ACKNOWLEDGMENTS

Firstly, I want to say a big thank you to everyone who helped me bring this book to life! From giving advice to a new author to helping with research, creating awesome illustrations, and helping me through this process, so many people had a part to play in getting this book into your hands.

In no particular order, a huge thank you to:
Anna Hands
Hazel King
Mercedes Beaudoin
Eliza Swanson
Charlotte Pointing
Kerry Maule
Isaias Hernandez
Robbie Lockie
Alexander Rigby
Lindsay Dobbs
Micah Schmidt
Cindy Kang
Alexander Woodrow
everyone at DK who worked on this book
and lastly, thank you Mum and Dad!

Secondly, I want to say thanks to you, the person holding this book, for caring about our planet and for being part of the movement to help create a safe and inclusive future for the following generations.